Sustainable Rural and Urban Ecosystems: Design, Implementation and Operation

Gunther Geller • Detlef Glücklich
Editors

A. Otte • R. Perfler • L. Richard • D. Simmering
Contributors

Sustainable Rural and Urban Ecosystems: Design, Implementation and Operation

Manual for Practice and Study

Springer

SPONSORED BY THE
Federal Ministry of Education and Research

Supported by:
Federal Ministry for the Environment, Nature Conservation and Nuclear Safety

based on a decision of the Parliament of the Federal Republic of Germany

Editors
Gunther Geller
Ingenieurökologische Vereinigung IÖV
Augsburg
Germany

Detlef Glücklich
Bureau for Sustainable Design
Buchholz Holm-Seppensen
Germany

First published by the Ecological Engineering Society IÖV
PO Box 102229, D-86150 Augsburg, Germany;
www.ioev.de; info@ioev.de

This publication is based on research, experiences and findings of two projects at the biggest private university in Ghana: Valley View University (VVU), Accra. The first research and development program was about ecological cycles at VVU, financed by the German Federal Ministry of Education and Research (BMBF), the second dealt with promoting VVU as climate friendly university and was financed in the Frame of the Climate Change Initiative of the Federal Government of Germany by the German Federal Ministry for the Environment, Nature Conservation and Nuclear Safety (BMU).

ISBN 978-3-642-28260-7 ISBN 978-3-642-28261-4 (eBook)
DOI 10.1007/978-3-642-28261-4
Springer Heidelberg New York Dordrecht London

Library of Congress Control Number: 2012940252

Printed on acid-free paper

Springer is part of Springer Science+Business Media (www.springer.com)

Contents

Chapter 1
Introduction

1.1 Background

These days human beings have an incredible influence on the planetary ecosystem, on climate change, on biodiversity, to name only two. Many of the ecosystems on the planet nowadays are more or less dominated by humans and that is true especially for the ecosystems, where men are living in, rural and urban settlements. Therefore it is most important to design and operate this kind of systems in a way that is really sustainable, which means long-term-functioning in harmony with the planetary system and human and non-human its inhabitants.

How to design and operate elements of the human ecosystems like houses, bridges, roads, sewer lines, factories is known to engineers, architects, managers. There is some understanding about ecosystems, especially natural ones and partly about their design and operation. However there is a deficiency in understanding rural and urban settlements as human ecosystems in a holistic way and nearly no experience in designing and operating them in a holistic and sustainable way.

This is especially true for the area of information, its understanding and application in rural and urban settlements. Some of the aspects being acceptance, commitment, decision-making, quality management, organizational structures, socio-cultural background, training and education, to name only some few. To try to close this gap this manual gives an explanation of the general principles of this rural and urban ecosystems and some holistic concepts and provides some methods, procedures and tools for their design, construction and operation. The experiences it is utilising are based principally on a practical showcase, namely the ecological development of Valley View University (VVU), the biggest private university in Ghana, and its campus.

The general approach of ecological engineering and quality management is applied and modified for settlement-ecosystem projects. Thus it encompasses all the steps of realization, from first discussions to operation and maintenance. In detail it consists of preliminary discussions in a prephase, via survey and inventory-taking to first concepts and the following decisions for the procedure and solutions.

G. Geller and D. Glücklich (eds.), *Sustainable Rural and Urban Ecosystems: Design, Implementation and Operation*, DOI 10.1007/978-3-642-28261-4_1,

The next steps then would encompass the design and the approval and then the award of contract and the construction or implementation. The last but not least step will be the operation and maintenance of the ecosystem.

For all the various steps various stakeholders are affected: for the prephase the citizens and representatives of the local authorities, for operation and maintenance the operational staff and units for example. Each again needs a different approach and different tools, some of which are included in the manual.

The informations, recommendations and tools are intended for the following target groups among others:

- Local authorities (giving hints for the procedure and the involved stakeholders)
- Designers (holistic approach, procedures, tools)
- Regulatory bodies, permitting and financing authorities (requirements for approach and procedures)
- Constructing and implementing firms and institutions (recommendations, tools)
- Operating bodies (hints for operation, tools).

1.2 The Example of Valley View University

VVU is situated in the outskirts of Accra, the capital, in a peri-urban and also still partly rural area. Thus it can serve as an example of an ecological settlement-ecosystem, which could be multiplied both in more rural settings (as eco-villages) and in more urban settings as well (eco-cities, ecological city-quarters).

In its development the VVU campus as an ecosystem has been adapted to the needs of its inhabitants and those of its ecosystem as well as possible. Because many of the needs do deal with problems and necessities, that are quite common all over Africa, like lack of water, electricity, trained people etc., many of the solutions found at VVU also can serve for Africa generally. VVU has another advantage: it is a unit of education, which can learn from its own ecological development and installations, can utilize this installations for education and training and finally can broadcast this existing positive showcase far beyond VVU into Ghana and Africa, also via its alumni.

Valley View University is the biggest private university in Ghana, and the first, which was accredited by the authorities. It started in the 1990s of the last century and 2001 it still was in its starting phase, with only few buildings and infrastructure existing and around 700 people on campus. Water was delivered by the VVU-own truck, ecological installations had not been put in place. Thus it offered a unique opportunity of developing a sustainable campus.

The advantages it brought in have been:

- Private institution with fast decision-making processes,
- A president and vice-president being in favor of ecological solutions,
- A campus of 120 ha including a lot of agricultural area,
- Not far from the capital and its airport,
- Ghana being one of the very few African countries with stable and peaceful conditions.

2003 it was possible to get the financing of a research & development program about ecological water solutions of the German Federal Ministry of Education and Research (BMBF). This enabled bringing ecological cycles of waters into operation in the course of the next years, all as part of an ambitious holistic ecological concept, that included an ecological masterplan and even the intention to start a study of ecological engineering as part of a new faculty.

2008 based on the success of the still running program it was possible to get financial means of the Climate Change Initiative of the German Federal Government via the German Federal Ministry for the Environment, Nature Conservation and Nuclear Safety (BMU), intended for a center for ecological studies, big volume rainwater storage underneath roads, planting thousands of trees, conserving existing old trees and natural vegetation and water treatment.

The common goal of developing VVU as the first eco-university of Africa has been identified in the course of the close working together and of a retreat, where 40 persons representing the most important sectors and stakeholders have been cooperating to positively let this vision emerse and the design and the next and most important measures jointly be agreed upon.

Now at the end of these two programs the experiences are evaluated and the recommendations for future ecological projects are offered in a way that practitioners can use and apply them, practitioners that always have too little time.

1.3 How to Use the Manual

The manual is intended to help the practitioners, which are dealing with human based rural and urban settlement-ecosystems in some of the steps of their realization (design, implementation, operation) and which are concerned about ensuring their sustainability practically. Thus it not only explains the general concept, the steps of realization and the respective involved stakeholders, but also gives hints and tools for the practitioner. Valley View University serves as the showcase, illustrating how to adopt the given information to other projects.

In Chap. 2 more explanations about human ecosystems and their components and characteristics are given.

Chapter 3 explains the key concepts and methods and goes into more detail for information and quality management.

In Chap. 4 Valley View University serves as real-case example, where the holistic approach was tested and is presented in detail and for the specific areas of Ecological Building, Land and Vegetation Management, Peri-Urban Agriculture, Water Management, Mass Flow Management and Information Management. For this fields all steps are elaborated, namely the development, design, implementation, operation and some economic aspects.

Chapter 5 summarizes the lessons learned.

Chapter 6 lists the involved authors, partners and institutions, in Chap. 7 some relevant literature is quoted.

Chapter 8 gives some tools, developed and tested and found useful at Valley View University, which may serve as an example for doing future projects that deal with human ecosystems in a more holistic and sustainable way. And Chap. 9 as Annex includes more examples of tools and applications.

Generally the example of Valley View University is just that: an example, which for the specific projects, circumstances, stakeholders, steps has to be adapted. This is also true for the tools, which shall present an easy-to-use framework. These tools are available as word-documents on CD too, for easy adaptation and use.

This manual is to be seen under construction like a loose-leaf catalogue, which shall grow in the future and shall encompass more test-cases and realized examples of human settlements as ecosystems. Thus if you want to contribute or want to make suggestions for improvement, please feel free to contact the Ecological Engineering Society IÖV.

Chapter 2
Human Ecosystems

2.1 Ecosystems in General

Ecosystems are systems, consisting of living beings, interrelated among themselves and with their environment (other ecosystems including the planetary ecosystem Gaia).

Ecosystems operate at a specific place, termed ecotop. The relations or fluxes in the ecosystem can be of energy, mass or information. Energy always is one-way with decreasing capacity for available working-power, the last step being low warmth without specific use for human use. Mass or matter in mature natural systems goes in cycles, most of it locally. Information is the relation, that controls the systems and the being's relationships.

The living beings according to their function in the system can be producers (like green plants, producing oxygen and nutrition), consumers (animals including man) and reducers (organisms that reduce materials to make them available again for the producers).

2.2 Human Ecosystems

The new thing about human ecosystems is that they contain elements and relations/ fluxes of a technical-cultural type (Fig. 2.1).

Technical-cultural producers are factories for goods for example, technical-cultural consumers are mass or energy-consuming things like refrigerators, machines, cars etc., technical-cultural reducers are wastewater-treatment plants, biogas-plants or composting-sites. Technical energy can be electricity, technical matter maybe produced goods like cars, technical information shows up as internet and telephone etc.

G. Geller and D. Glücklich (eds.), *Sustainable Rural and Urban Ecosystems: Design, Implementation and Operation*, DOI 10.1007/978-3-642-28261-4_2,

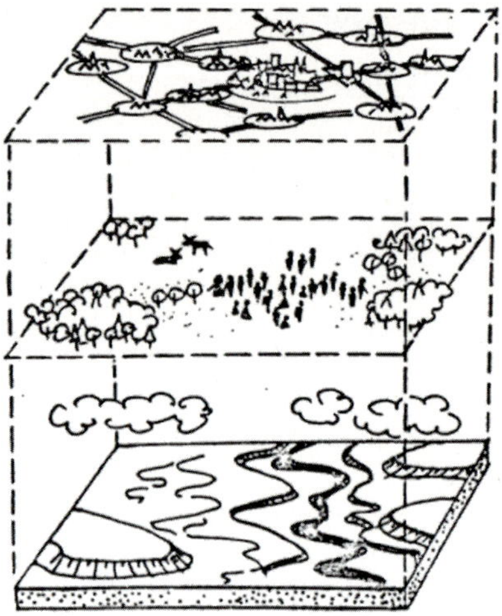

Fig. 2.1 Levels of ecosystems (technical-cultural components on *top*, living elements in the *middle*, inanimate parts on the *bottom*) (Tomåsek 1979)

2.3 Development of Ecosystems

Ecosystems generally develop in space and time.

Development in time, also termed succession, means proceeding from an unripe state to a mature one. In the pioneering first phase ecosystems only have few elements and few connections between the elements and their environment. After many steps in the course of evolution in time they can reach a ripe mature stage, where many diverse elements very intensely are connected by fluxes of information, energy and matter, utilizing the available matter, energy and information at its very best.

A measure of this mature stage is the very small loss of matter; i.e. erosion being of small importance. The same applies to natural and human-made systems likewise. In the final ripe stage, import and export of matter and (technical) energy is small and many diverse elements are connected very intensely and harmoniously.

There are characteristics of mature ecosystems, that can serve as a general orientation for sustainable human ecosystems too:

- Effective use of energy
- High biodiversity
- Closed cycles of minerals
- Low entropy
- High stability
- Self-organisation
- Sophisticated cooperation etc. (see Odum 1971).

To ensure that the human ecosystem takes the course of sustainability, e.g. of a mature system, during the whole development care has to be given to the amount of maturity = quality gained and to improve it if necessary. In quality management it is called circle of quality. Process management very much means management of information, including among others training, education, public affairs and acceptance inside and outside. The structure of the system in the end of the process must allow this mature ecological system.

The development of a human ecosystem in time makes it compulsory to accompany this process over the whole span of realization, from the first discussions through the planning phase till its final envisioned stage by a lot of measures like maintenance, training, modifying the hierarchy if necessary etc. These measures all serve the aim of gaining the final mature state of the ecological development in ecosystems. The development in time should be in a way that the system is able to adapt to it without loosing its balance.

2.4 Relations and Fluxes in Human Ecosystems

There are three types of relations or fluxes in an ecosystem: mass or matter, energy and information. It is important to note that these can have the form of flows or relations, but also that of a structure. Information for example can be in the form of communication (flow) or show up in the form of the hierarchical structure of a university or community.

2.4.1 Matter/Mass Flow

For ecosystems the type of mass flow can give a good hint about the maturity of the system. If there is a lot of cycles and little input and output, the system is autarc to some extent. If there is a lot of output, for example as erosion of soil, the system is not sustainable at all.

For the mass flows a balance is helpful for designing a sustainable ecosystem. One of the flows easiest to evaluate and model is water in its various forms.

2.4.2 Energy

Energy is a one-way flow: once used it looses its ability to do the same quality service again. In the worst case as low-temperature heat it is lost for human use. In natural ecosystems therefore energy is used on as many steps as possible and as effectively as possible.

The most important energy, stemming from our sun, comes in a non-material form. It is the most important renewable energy too. In human ecosystems however, energy often is delivered in a material form as oil and coal and wood, thus contributing to the mass flow balance too.

2.4.3 *Information*

In ecosystems, information is one of the three fluxes (matter and energy being the other two). While it is not the most visible one, it is nevertheless a flux of utmost importance. It is the flux, by which the system is controlled, its course steered and by which all the connections of the various members are started and maintained. Information in ecosystems can take on various forms. Generally there can be types predominant in natural systems, like informing partners about sources of food. In the technical-cultural human ecosystems like villages and cities, the technical-cultural type of information is prevalent. One of the more technical forms is the Internet and telephone connections. More cultural forms are values, ethics, beliefs, rituals, education, intercultural communication, etc.

2.4.3.1 Information as Socio-Cultural Framework

In technical-cultural ecosystems, controlling and managing the ecosystems is partly done by religious and socio-cultural sets of beliefs, mindsets, visions, paradigms, values, ethics, etc. Our belief in the necessity of continuous growth is one of these sets.

Without understanding those kinds of mechanisms, the socio-cultural background of the human system especially, measures from inside and outside will not have the intended long-term effects. This is crucial especially, when people from other cultures are operating in strange socio-cultural frameworks, like in the development business. Thus before and parallel to "doing something" the gaining of understanding of the specific framework first is a must and precondition.

The degree to which the specific human ecosystem will fit in with the global ecosystem, and whether or not the specific ecosystem and its human members will act in accordance with ecological principles, will depend very much on the extent to which ethics and values of the human members emphasize this ecological core.

2.4.3.2 Information as Organisation

Information can not only appear in the form of flux, but also in its structural form, the (hierarchical) organisation of human ecosystems like companies, communities like cities, villages and universities for example. This form of information is especially important in respect to the long-term functioning of the system, i.e. its sustainability.

In case of human ecosystems, one part of the ecosystem is the hierarchical organisation of the community or university itself. The structure of this system very much defines the flow of information inside and outside, the distribution of responsibilities, the general acceptance of people and measures and the availability of financial resources, to name only a few factors.

Thus for the ecological development of human ecosystems it is necessary to have positions on all level of hierarchy responsible for the ecological development and filled with appropriate persons. This person must be capable, trained, willing and given the necessary authority to deal with their tasks.

2.4.3.3 Information as Education, Training and Studies

To gain the necessary awareness and acceptance concerning ecological matters education, training and studies are essential means.

Depending on the kind of human ecosystem we have to deal with, different kinds of education, training and studies may be important. Studies are important for universities especially. For villages and towns specific trainings and educational programs for the inhabitants are essential. This can be empowering specific groups of the community, like women, or the youth members, or do training for specific skills, like operating a biogas plant, composting toilets, solar systems or water cycles.

2.4.3.4 Information as Communication, Acceptance and Public Relations

Human ecosystems will be functional in the long term only, when there is full acceptance inside and outside. To achieve that, distributing information will be necessary, but in itself will not be enough.

It will be necessary to spend time with the stakeholders and to provide possibilities for building up trust and working relationships and also to provide a framework where the involved can grow personally. Albeit its not possible to ensure acceptance and deep communication by measures for sure, however it is possible to provide a favourable environment for that.

Some suitable approaches and methods are: Appreciative Inquiry (Ludema et al. 2003), U-Theory (Scharmer 2007), The-Work-That-Reconnects (Macy & Brown 1999), Circle (Baldwin 1998), World Café (Brown & Isaacs 2005), Music and Dance etc. Examples of this approaches and methods applied at VVU are given later.

2.5 Engineering of Ecosystems

The combination of the holistic way of understanding in ecology and the practical approach in engineering brought forward a new approach, termed "ecological engineering".

Ecological engineering is existing now since the 60s, with a broad scope of projects and experience.

Ecological engineering could be defined as:

- "Environmental manipulation by man using small amounts of supplementary energy to control systems in which the main energy drives are still coming from natural sources" (Odum 1962) and
- "The design of sustainable ecosystems that integrate human society with its natural environment for the benefit of both" *(Mitsch 1996).*

The main tools of the ecological engineer are ecosystems and their components. Till now, the main focus was on ecosystems, influenced by humans (like rivers) or even completely designed or constructed by humans (like Constructed Wetlands). Human settlements however very rarely have been dealt with as ecosystems, designed, realized and operated according to the principles of natural ecosystems and with all of the components of the human ecosystem as tools.

In the VVU-example this approach of ecological engineering has been tested with the campus of VVU. This manual contains the summary and the practical approach and tools developed thereby.

Chapter 3
Design, Implementation and Operation of Human Ecosystem: Concepts and Methods

3.1 The Holistic Approach: Key Concepts and Methods

3.1.1 Introduction

For designing, constructing and operating human ecosystems practical approaches and tools are required. The general requirement is that they must be based on a holistic ecosystem approach, which encompasses all the ecosystem elements: the place, the fluxes and the inhabitants. And because it is a human ecosystem, it must include the technical-cultural parts of the system in particular (Fig. 3.1).

The place or ecotop includes natural aspects like geology, soils, flora (vegetation) and fauna, waters below and above surface, but also the cultural-technical ones like buildings, streets, infrastructure, structure of the community, the socio-cultural background etc.

Ecosystems develop in time and space and according to some fundamental principles or ecosystem laws. The ecosystem is something not or not fully existing in the beginning and there have to be made the right steps in the right order (i.e. the right process has to be undertaken), to reach the desired mature state of the ecosystem. Concerning the development in space means, that the available place has to be used in a proper way, hereby ensuring the harmonious functioning of the parts of the ecosystem. That also makes it necessary that all the functions of the specific system have to be kept in mind simultaneously and that not only one function is maximized on the expense of others.

The features, that comprise a holistic vision of such an eco-settlement or ecocity, are manifold and are illustrated in the Fig. 3.2

For planning ecosystems, it is important furthermore to have in mind, that the various elements and functions are interrelated and thus should be delt with not only separately, but also in their interdependences. An example concerning energy and material flows and an approach to handle that gives the following Fig. 3.3

To deal with all these aspects a general orientation or principle (Leitbild) will be necessary, where the process and the final general masterplan for the spatial planning

G. Geller and D. Glücklich (eds.), *Sustainable Rural and Urban Ecosystems: Design, Implementation and Operation*, DOI 10.1007/978-3-642-28261-4_3,

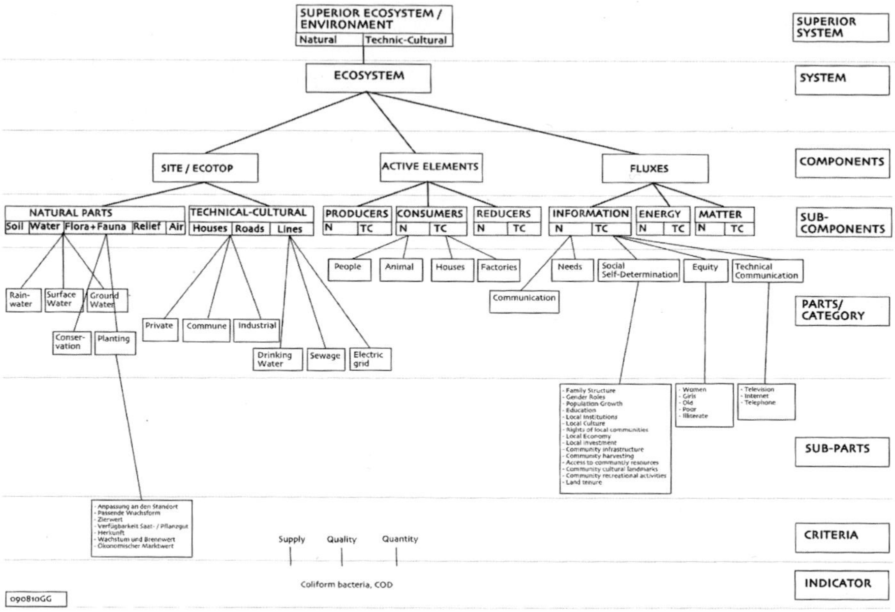

Fig. 3.1 Ecosystem overview: levels and parts

is oriented to. For the ecological development of human ecosystems the townscape or Stadt-schaft principle will serve this purpose.

Some of the concepts and approaches of ecosystem engineering of settlements are explained in more detail in the following, most have been tested (and found useful) at VVU.

3.1.2 Common Aim "Stadtschaft" or "Townscape"

Such a holistic approach must consider a lot of numbers of 1 parameters. In combination countless solutions prove combination nation. The system is not easy to understand completely. If not the essential aspects are considered, the solution becomes faulty. Therefore, a method is to be found after which systematically useful solutions can be worked out and be discussed.

Up to now the man-made landscape was formed under many single points of view, aims were pursued at most in a few points particularly in the energy area or nature conservation, an interlinking of both nearly unknown. Eco systems form is more and needs a leading goal, so a common line of view, a problem definition. Such a leading aim must be concretised by partial aims. Aims can be reached with certain measures or not, however, it should be checkable and must be declared again clearly. The creation of ecological systems becomes clearer, can be discussed, means for and ways for solutions can be improved step by step.

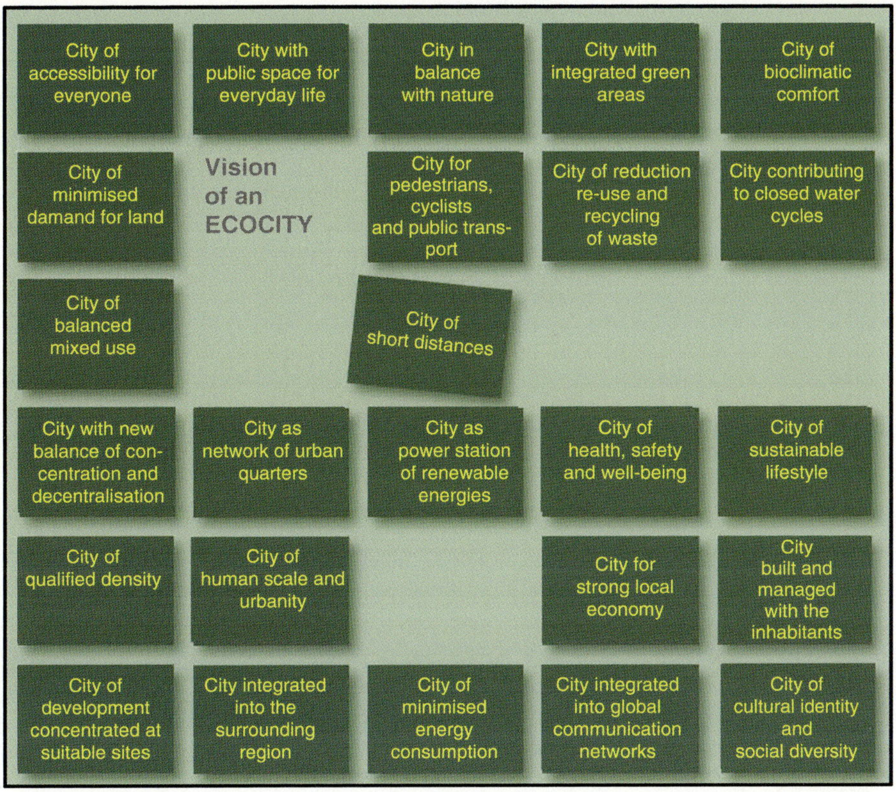

Fig. 3.2 Features of an ecocity-vision (Gaffron et al. 2008, 10)

Towns and her surrounding are an especially important life space, his sustainable creation is looked by many as a key for the survival of the people. Newer investigations confirm already up: By the bioactivity which depends in the essentials on the water availability an important and inalienable leading size must also be in creation of the towns. Because water is lives inalienably for ours daily, the lasting water cycle must be a basis of any man-made landscape and with it also of the town.

In the ideal case the sustainable town is also a stable ecological system like every sustainable landscape—town and landscape in one system, we call it "Stadtschaft" (Townscape) as a part of the whole ecological system: "Stadtschaft" is leading aim. The ideal case will be hardly reached, but the aim give a wide frame for activities: High bioactivity with efficient water and material use, efficient energy use without negative long-term environmental impacts, high interest in local physical and human resources (we call it a high base quality)—however, no autarchy; since only by interlinking of the Stadtschaft with the landscape and other resources synergies are used for a good price after the motto: Not everybody is able to do everything.

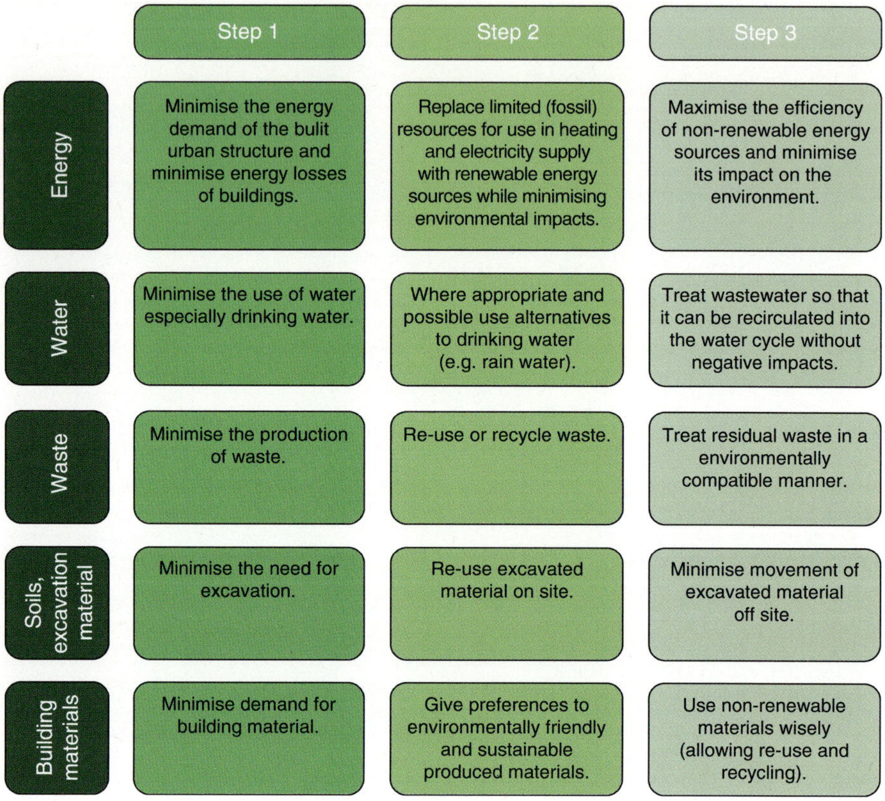

Fig. 3.3 Three-step-strategy for energy and material flows (Gaffron et al. 2008, 32)

Work in partial steps with partial aims: To realise the Stadtschaft in one step is not possible, the system is too complicated. So we have to go on step by step—concerning the partial fields, the function and the space. A cell model was developed as method to work systematically.

3.1.3 *Cell Model: Functional and Spatial Units*

"Stadtschaft" is formed by many single parameters. Though concepts like garden city, green town, solar town' etc. reflect images, however, they are too vague for the concrete work with concepts. Simple estimates cannot lead in the complicated system to a sensible solution. Therefore, the small units are created for investigations. According to kind and size it can be the e.g. building of the same use, parts of the open space with the same character or a decentralised sewage disposal of sewage of the same kind. These small functional units are called according to the model of the living beings cell. They correspond to the principle of the partly decentralisation and are functional and spatial units and action rooms in which the planner develop Stadtschaft.

The cells are no self-sufficient units. They have mostly connections to their surroundings and need it. Cells are closed in parts, in others again openly, as this are, e.g., a building and town accommodations by her connection to the surroundings. The function of a cell and her connection needs a constant review. Cells can be coupled with each other and form thus bigger functional units (cell groups).

The concept of the cell calls not only a pure model of thought ("mental cell"), but is a working model for practicable decentralised planning and design in spatial and functional regard. By the combination of cells in cell groups und again cell groups to bigger units complex structures can be built up.

The whole campus area of the VVU could be a cell what would have simplified planning and realisation certainly. Nevertheless, this contradicts the principle of the decentralisation and the higher leading aim of the bioactive Stadtschaft. Often central solutions are technology-intensive solutions which cannot react adaptably to special uses or changed requirements. Establishing cycles in centralised solutions mostly would mean costly and long ways. The advantage of decentralised solutions is that optimum solutions can be found for the different cells and come with it most near the general principle of a bioactive Stadtschaft.

3.2 Integrated Management of Process, Information, Quality

3.2.1 Generally

From the ecosystem perspective, information management and quality management both belong to the fluxes of information and their handling. Quality management always is understood as an ongoing cycle of improvement, thus the process aspect is included generally (Fig. 3.4).

Where information-management mainly is seen from the angle of availability and distribution of information and its impact on the stakeholders (resulting in commitment for example), quality-management more deals with the process as such and the quality of its outcome in respect to the users/stakeholders.

3.2.2 Management of Information

Information is the key for the handling of human ecosystems. It must be management in a way, that all necessary information is available and that all stakeholders can and will contribute to this availability.

It is relatively easily possible to make information available: via power-point presentations, via emails, via Internet, via meetings etc. The effectiveness however is not so much a question of the amount of available information and its distribution,

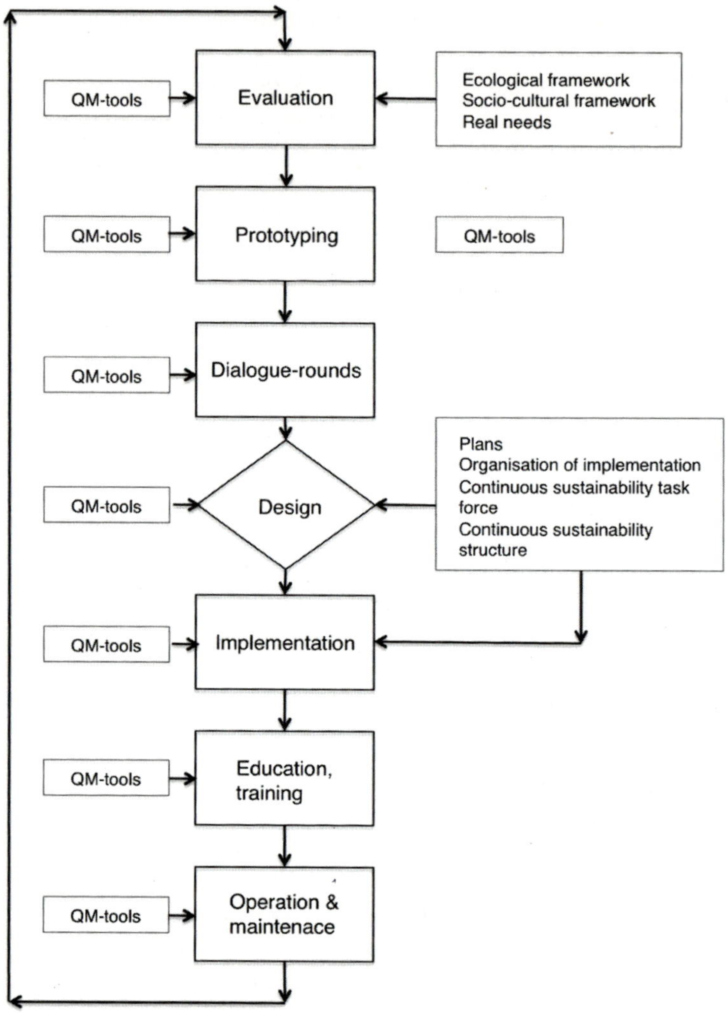

Fig. 3.4 Integrated management of process, information and quality

but is mainly concerning the quality of information and depth of understanding (and listening) and lately the commitment (Fig. 3.5).

This makes a different kind of approach necessary, which goes far and deep beyond the surface and the usual ways of communication. It must find out the real needs of the stakeholders, include them from the start and throughout the whole process of realization and must enable communication and understanding from heart to heart and spirit to spirit. Without that long-term commitment and via that long-functioning of the system (sustainability) cannot be achieved.

There are approaches available and even tested and applied successfully in various organizations. One of these is Presencing, a method developed at the Massachusetts

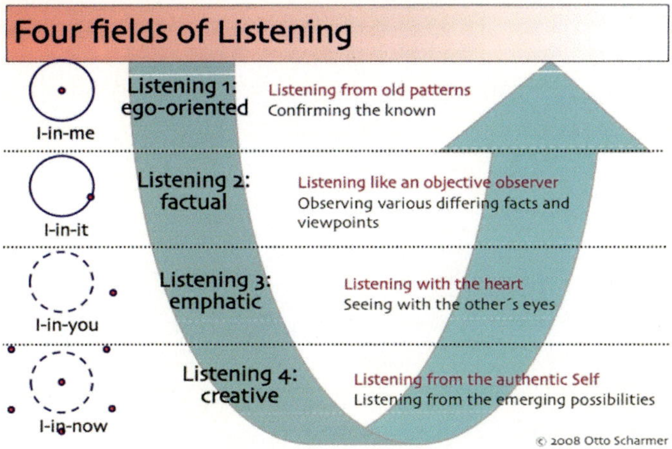

Fig. 3.5 U-Process: Illustrated for the field of listening (Scharmer 2007)

Institute of Technology (MIT) (Scharmer 2007) and which for Listening is illustrated in the figure above.

3.2.3 Management of Quality (QM)

Whereas information management normally is understood in a general way, quality management more serves the purpose of gaining the satisfaction of the customers. The satisfaction is realized by delivering product, that fulfill the requirements of the customer in the most economic way. Quality management is understood as a process of continual improvement (Fig. 3.6).

Originally Quality Management was intended for production processes, later it also was applied on other fields like services and engineering. For our holistic approach we apply it on all the various stakeholders in the process and also on all steps of the process. We use QM-tools and modify them to the needs of a holistic approach for human ecosystems.

3.2.3.1 The Steps of the Process

In projects of establishing human ecosystems, the whole process consists of the following main steps:

- Prephase
- Design
- Implementation
- Operation and Maintenance

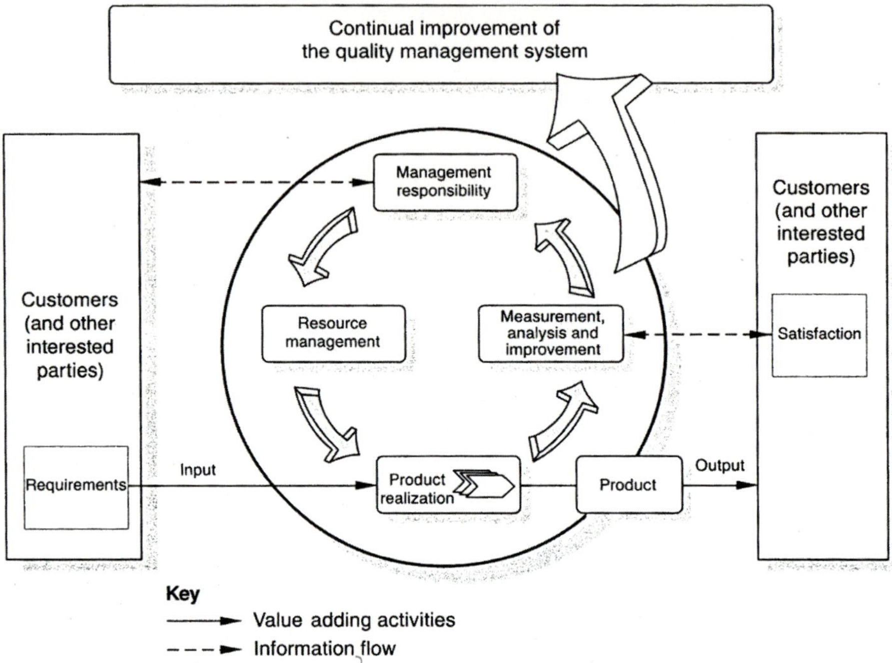

Fig. 3.6 Cycle of quality management DIN/ISO 9001

These main steps usually consist of various sub-steps, depending on the kind and complexity of the project. For the design-step normally first concepts or sketches are followed by more detailed solutions, which finally have to be brought in a form that enables realization or implementation.

3.2.3.2 Contributing Actors

The steps of the process are executed by varying actors and by including varying stakeholders. For the step of design for example the client or contracting body, authorities and the designers, engineers, architects are important, whereas for the step of operation the operating staff is essential,

3.2.3.3 Tools

To make sure that the process of realization is carried out in the best possible way concerning its holistic and sustainable quality, tools are helpful that detail the various steps, their input and output (results) and the respective stakeholders. QM-tools like process-descriptions, flow-charts, checklists, forms, manuals, etc. can serve as input too.

3.2.4 Integrated Management of Process, Information and Quality: Procedures and Tools

3.2.4.1 Action Research

For all ecological projects and their continuous improvement process it is essential to put the state of affairs in the beginning on record and again at the end. This not only is an element of quality management, but also a determining one of an approach, called action research.

This is an elegant combination of at the same time doing, learning, researching and improving. It was developed in the social sciences first, but is used in many other fields now. Action research is an interactive inquiry process that balances problem solving actions implemented in a collaborative context with data-driven collaborative analysis or research to understand underlying causes enabling future predictions about personal and organizational change (Reason and Bradbury 2005).

Normally this kind of approaches are separate: in a research and development program you are doing research and stay distant, not interfering in the system (at least you try it), whereas in an project of implementation or investment you do something without taking too much care of the process, the changes in the realization of the project, what all participants learned and what could be concluded for future projects.

Action research allows for taking part in the change of the system, but records this changes, the situation before and after and the things being learned. The effort can be small and generally for practical projects must be. The tools can be lists, investigations, protocols and others. For a workshop questionnaire before and after would be part of the action research. For application in development cooperation see Gagel 1995.

3.2.4.2 Action Learning

Action learning (following definition see Wikipedia) focuses on research into action taken and knowledge emerges as a result that should lead to the improvement of skills and performance.

If the main focus is on participatory learning, it is termed Action Learning. It means an educational process whereby the participants study their own actions and experiences in order to improve performance. This concept is close to learning-by-doing and teaching through examples and repetitions. Action learning is done in conjunction with others, in small groups called action learning sets or two-in, two-out team. It is proposed as particularly suitable for adults, as it enables each person to reflect on and review the action they have taken and the learning points arising. This should then guide future action and improve performance.

The method stands in contrast with the traditional teaching methods that focus on the presentation of knowledge and skills (see Kramer 2007).

3.2.4.3 Appreciative Inquiry (AI)

This is an approach for the development of organizations, that is based on the idea, to build organizations around what works, rather than trying to fix what doesn't. It is the opposite of problem solving. The question: what is the best we can be is in the foreground and as a result the process of improvement does go on permanently (like in quality management). Whitney and Trosten-Bloom (2003) define AI as *the study and exploration of what gives life to human systems, at their best.*

Appreciative Inquiry is a particular way of asking questions and envisioning the future that fosters positive relationships and builds on the basic goodness in a person, a situation, or an organization. In so doing, it enhances a system's capacity for collaboration and change (Wikipedia).

Appreciative Inquiry utilizes a cycle of four steps:

- DISCOVER: The identification of organizational processes that work well.
- DREAM: The envisioning of processes that would work well in the future.
- DESIGN: Planning and prioritizing processes that would work well.
- DESTINY: The implementation of the proposed design.

The approach of Appreciative Inquiry can be seen as follows:

- AI sees organisations as living spiritual-social systems.
- AI works in the affirmative, continually seeking to discover what gives life.
- AI encourages the stakeholders to be facilitators of possibilities, hope and inspired action and
- To continually seeking ways to support others of making it their own.

3.2.4.4 The Work that Reconnects (TWR) (Joanna Macy)

This approach was developed by Joanna Macy and co-workers and can be seen as a spiral. The first starting stage is Gratitude, that is followed by "Honouring our pain for the world". After the stage of "seeing with new eyes" the step of "Going Forth" ends the first turn of the spiral, which then proceeds with this ongoing process (Fig. 3.7).

TWR synthesizes approaches of ecology, system theory and Buddhism to perfectly fit the ecological requirements of today.

Its uniqueness very much comes from the stage "Honouring our pain for the world", which allows for an expression of all the suffering of the participants. This suffering is understood as coming from the interrelatedness of all beings in the web of life that cannot be reduced to a simply individual psychological topic. This integrated expression of the sorrows in this step frees the energies for the next stages and simultaneously leads to a deep connectedness of the participants. For all the stages a lot of special exercises are available (see Macy and Brown 2003).

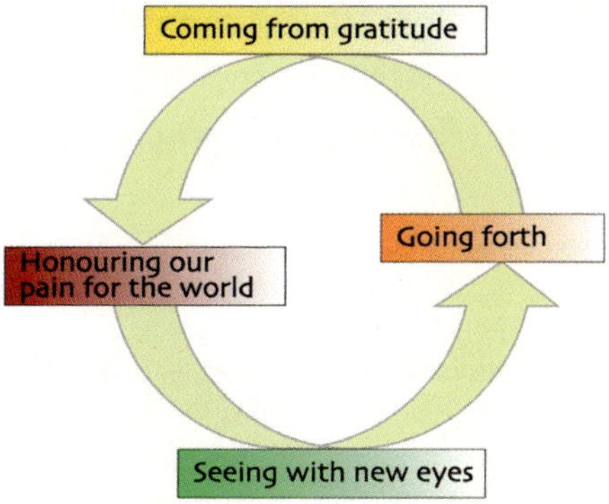

Fig. 3.7 The spiral of "The work that reconnects"

3.2.4.5 U-Theory

"Presencing," a blend of the words "presence" and "sensing," refers to the ability to sense and bring into the present one's highest future potential—as an individual and as a group.

Theory U offers both a new theoretical perspective and a practical social technology. As a theoretical perspective, Theory U suggests that the way in which we attend to a situation determines how a situation unfolds: *I attend this way, therefore it emerges that way.* As a practical social technology, Theory U offers a set of principles and practices for collectively creating the future that wants to emerge (following the movements of co-initiating, co-sensing, co-inspiring, co-creating, and co-evolving) (http://www.presencing.com/presencing-theoryu).

The process can be seen in the form of an U, where the left start of the U is the initial part of the process, and the one, being on the most shallow surface level, the bottom is for the deepest, most inspiring phase, where the emerging future can be sensed and the right wing of the U stands for the steps of realization (see Scharmer 2007).

3.2.4.6 Circle

Exchanging in circles is an age-old practice, without interruption used of indigenous people and nowadays also used of various groups and purposes. Always it is used to get commitment and consensus. The general attitude is speaking and listening with intention and self-monitoring one's impact and contributions. It is based on the principles of confidentiality, shared responsibility, rotating leadership and relying on the spirit. The one speaking gets the full attention and respect of the others, which do not interrupt and discuss, but simply listen. When the speaker has ended, it's the turn for the next in the round, again without discussion interfering. The native Americans used a talking stick, the one holding it having the authority of speaking until he delivers to the next (see Baldwin 1998).

3.3 Vegetation Management

3.3.1 Ecosystem Functions and Services

Scientifically defined, vegetation is the entirety of plant species assemblages at a certain site or in a certain area. For reasons of simplicity, however, we consider here all vegetated space, i.e. "green" land cover or "green space", as vegetation. Vegetation plays without doubt a crucial role in the functioning of human-dominated ecosystems and provides key benefits (ecosystem services) to the human population. The most fundamental of these "ecosystem services" are the production of biomass as food and energy resource and the release of oxygen. But in human ecosystems, a well-developed plant cover also reduces heat, noise, dust, and serves as wind break. It further increases water infiltration and reduces runoff, thus facilitates flood control and groundwater recharge. Besides these more physical properties of vegetation and their effects on ecosystems, a quality green environment has numerous psychosocial benefits and promotes human wellbeing and health.

Every type of vegetation serves also as habitat for plant and animal species and is thus the foundation of biodiversity. This applies not only to natural or semi-natural ecosystems or landscapes with limited or moderate human impact, like e.g., forest reserves or traditional agricultural landscapes. It is also true for rural and urban environments. Many studies have shown that the number of bird and plant species can be higher in urban green space than in the surrounding landscapes (cf. Marzluff et al. 2008). Hence, the conservation of biodiversity and a wildlife-friendly management is an important issue in rural and urban systems.

3.3.1.1 Guidelines in Vegetation Management

There is a growing awareness for the invaluableness of a well-managed and designed green space for the sustainability of urban and rural systems (Greenkeys Team 2008). However, this is not at all reflected by the current trends. Rapid urbanization across the world causes losses of green space in urban and peri-urban areas at alarming rates (Gairola and Noresah 2010). Being a public good, it is crucial that green spaces and vegetation management should receive more attention in urban planning and development. Particularly for ecological and holistic planning approaches, it is important to not only look at the amenities the green space provides for the human population. The functioning of urban and rural green as wildlife habitats has to be given more weight. For a holistic approach in the design of human ecosystems, vegetation and land management should thus follow two guidelines:

- Vegetation management should aim to optimise the regulation functions (energy and nutrient exchanges, e.g., cooling, carbon and water storage) of the natural elements in the human ecosystem (e.g., soil, wetlands, forests).
- Vegetation management should secure the needs of the community while not compromising the needs of wildlife and biodiversity.

3.3.1.2 Types of Green Space

Certain areas of green space serve certain purposes. In most cases it will be reasonable to make a preliminary distinction between:

- (Nearly) natural and semi-natural habitats, with the main purpose to support wildlife
- Farmland and gardens, used mainly for the production of food and energy
- Public and private (urban) green space, used mainly for recreation, leisure and social activities

3.3.2 Conservation Value of Natural and Semi-Natural Habitats

Natural and semi-natural habitats are forests, shrubland and hedges, rough grazings, wetlands and streams, sources and shoreline areas, rocky outcrops, and all types of vegetation that are not strongly impacted by human activities. The priority should always be to conserve these habitats as far as possible and to plan for a sustainable management that considers the needs and requirements of the respective habitat type. To identify the conservation value and the management requirements, ecologists have to be involved to quantify, describe and assess the quality of the existing habitat types (Lovell and Johnston 2009, see below).

3.3.2.1 Landscape Context

All ecosystems interact with their neighbouring systems through the flow of air, water, fire, soil, and substances. They are affected, but they also affect their environment. This is especially true for habitat types and the movement of species living therein. These rely on the exchange of populations and the possibilities to move inside a habitat network. For the assessment of the conservation value of natural and semi-natural habitats found at a site, it is hence particularly important to consider the broader landscape context at various spatial scales. In addition to topography, hydrology, soil types and other biophysical landscape features taken into account for the ecological design of the human ecosystem, a firm knowledge and understanding of the distribution pattern of the respective habitat types (and of the plant and animals occurring in these) in the surrounding landscapes, regions, up to the national scale, is needed. Local and expert knowledge, literature research and (land-use) maps or satellite images are the basic sources of information at this stage. GIS-techniques are usually applied to process the data and to generate different layers of information.

Example
A remnant forest fragment of a few ha size in a project area may serve as a simple example to illustrate the relevance of landscape context. Although a forested area provides always numerous services for the whole system (water storage, clean air etc.), the conservation value of this forest fragment is highly dependent on the surrounding. It is particularly high, if the urban ecosystem is part of a large (peri-)urban agglomeration without any or few other remaining natural habitats. Due to its rarity in the area, it may be of fundamental importance as living space for woodland species occurring here, whereas the same forest fragment would have considerably less conservation value if it was located in a small town surrounded by large forested areas. However, most urban and peri-urban ecosystems lack sufficient areas of natural and semi-natural habitats due to the accelerating speed of urban sprawl.

3.3.2.2 Habitat Quality and Restoration

Habitat quality and restoration are other important aspects. In human ecosystems, the remaining natural and semi-natural habitats are often degraded by inappropriate management practices and excessive overexploitation, which is all too often accompanied by the invasion of unwanted, problematic species. Restoration techniques (sustainable management schemes, reforestation, seed transfer) can be used to improve the situation and restore habitat quality.

3.3.2.3 Habitat Networks

In addition to habitat area and quality, the functional significance needs to be taken into account. In rural settlements and farmland we often find networks of semi-natural hedgerows. The total area they occupy may be negligible, but their functional importance for the exchange and dispersal of species is immense. As a rule of thumb, linear habitats which are exposed to low or moderate anthropogenic impacts have a high significance for biodiversity conservation. They not only host a number of resident species but provide also shelter and connectivity for moving species in an otherwise hostile matrix of settlements or farmland. In urban environments, parks and/or networks of old home gardens in residential areas can potentially serve as an important habitat network, provided the habitat quality is sufficient.

3.3.3 *Farmland Management*

Farmland is agricultural land designated to grow either annual (cereals), perennial (e.g. banana, strawberry) or tree crops. Further, it includes land used to grow forage (meadows and pastures) and fallows. Complex, traditional agricultural landscapes provide many ecosystem services, feature large proportions of semi-natural, non-crop habitats and are important habitats for many species (Perfecto et al. 2009). Highly intensified industrial agriculture, in contrast, relies on a high input of nutrients and pesticides, which is increasingly compromising the quality of soil, groundwater and freshwater resources around the world. It results in simplified landscapes with large fields and a decreased proportion of non-crop habitats. In many landscapes, the introduction of high input agriculture has led to a decline in farmland biodiversity, with the remaining biodiversity concentrated in field edges and non-crop habitats (cf. Bianchi et al. 2006).

In rural ecosystems, farmland covers usually large areas. The urban consumption of farmland is one of the main fields of conflict in peri-urban development worldwide. A holistic planning for urban and rural ecosystems should thus aim at the preservation of agricultural land and promote low-input agriculture. Numerous studies have shown that organic farming reduces environmental risks in terms of pollution and resource degradation, and promotes farmland biodiversity.

3.3.4 Urban Habitats: Public and Private Green Space

As mentioned above, we can observe large numbers of plant and animal species in many urban ecosystems, even without a substantial area of natural habitats. One reason for this pattern is that urban areas host many species, which were either deliberately or incidentally introduced by man over the last few centuries. A larger number of species however has adapted to the habitats created by humans over the past hundreds and thousands of years. Others have immigrated from habitats and ranges that provide resources similar to urban and rural environments (e.g., rock-nesting Peregrine Falcons on top of sky-scrapers, grassland plant species in urban park meadows). Virtually all types of urban areas have thus a capacity to support wildlife and to serve as habitat. Ecologists therefore classify "urban habitat types".

3.3.5 Management matters

It is important to note that all farmland and urban habitat types have the capacity to support characteristic assemblages of plant and animal species. For a holistic approach it follows that this potential needs to be considered in the design and management of the human ecosystem. To this end, the individual management of the habitat determines if this potential is realised. Often, only little changes in management practices are needed to improve habitat quality without compromising human needs.

For farmland management we would propose organic farming, but even in modern agriculture a slight decrease of nutrient and pesticide inputs in field sections or along the edges can improve the situation (deintensification) for wildlife. The reestablishment of small semi-natural landscape elements in the farmland matrix (see habitat networks) is also an effective measure to promote farmland biodiversity. In urban green spaces, a reduction of mowing frequency in selected areas of a park or public greenery, the refraining from the use of pesticides, the preservation of old trees and shrubs, the planting of suitable trees and other measures will support wildlife (e.g. Lyle 1999).

3.3.6 Client and Stakeholder Involvement

Natural habitats, farmland and urban green space are usually used and managed by a number of different stakeholders. For development projects involving larger areas of vegetated space it is thus essential to include the relevant persons and groups of interests at an early stage of a project to assess and discuss land use conflicts. At the same time, the assessment of flora, fauna and vegetation has to be conducted. The results of this assessment define the range of design options presented to the client. These will be discussed in an interactive process to find an agreement on the main issues regarding a future sustainable land use.

3.3.7 Conclusion

The sustaining of ecosystem functions and services provided by vegetation is one of the main principles in ecological planning (Lyle 199). It is obvious that in the design of human ecosystems an adequate amount of space needs to be reserved for green space. The conservation of natural and semi-natural habitats is a priority issue, which needs to be discussed and evaluated in the landscape context. Hence, a thorough assessment of the natural resources including biodiversity has to be carried out by ecologists in an early phase of a development project (cf. Environmental Impact Assessment, see IAIA (2005, 2010) for more relevant literature). Farmland and urban habitats can be improved by a well-designed habitat network and a proper management. For the integrated planning of projects at a smaller spatial scale (object planning) it is also crucial to include a well-designed and managed green space following the ecological principles described above. To make projects sustainable, it is important to involve stakeholders and land owners at an early stage of a project.

3.4 Cycle-Management

3.4.1 Water Management

Water management of urban and rural settlements has four main objectives:

- Provide access to safe and sufficient water

- Provide access to safe and sufficient sanitation
- Ensure flood prevention
- Prevent water pollution

Water is required for domestic, agricultural and industrial purposes (drinking, cooking, flushing, washing, irrigating…). Therefore to sustain the community and enable its development, it is essential to supply enough water at an adequate quality to the community. The water quality required will depend on the type of use. For instance, drinking water has to be "safe" referring to microbial, physical and chemical aspects. It should contain limited concentrations of chemical substances, such as nitrate and should be free of harmful constituents, such as pathogenic microorganisms. Drinking water quality requirements are generally defined by e. g. guidelines from the World Health Organisation (WHO 2006) which also provides guidelines for safe use of wastewater, grey water and safe water for irrigation. The safety of water supply and sanitation is of major concern and should be assessed and managed in the frame of a water safety plan. On the base of such principle requirements local regulations should support practical implementation by taking into account the specific conditions on site.

The local and regional context has to be assessed regarding the availability, quantity and quality of potential water sources (surface water, groundwater, rainwater, reuse). The objective of providing safe and sufficient water to the community can be achieved in the short and long-term by implementing water management measures focussing on sustainability of resources, demand management and efficient use and reuse. Additionally, the assessment and development of public and consumer awareness, of organisational and institutional frame and of monitoring and backstopping schemes for operation and maintenance are essential requirements for successful implementation.

Besides the general objective of efficient water use, this becomes essential with water scarce environment. A key step toward the efficient supply of water is the monitoring of water quantity provided to the distribution system. Water meters at key locations of the distribution system and at consumer side can provide information for a "water balance" of supply and consumption. Additionally, it helps to identify water losses. The implementation of water saving "technologies" such as low flush toilets or water saving shower heads at consumer level is a key issue for the reduction of water demand and requires awareness-raising, good communication and information at consumer side. The acceptance of technologies by the community should however always be checked prior to any implementation.

For the well being of the community, not only a safe and reliable water supply is required but also a safe access to sanitation. Wastewater should be disposed to preserve public health. Poor sanitation practices are unpleasant and may lead to the outbreaks of severe diseases within the community. Wastewater disposal can be achieved by various strategies: e.g. the conventional centralized systems, which consists of a combination of a sewer and a wastewater treatment system. The wastewater is in this case usually considered as a waste which must be collected, treated and discharged into a receiving water body.

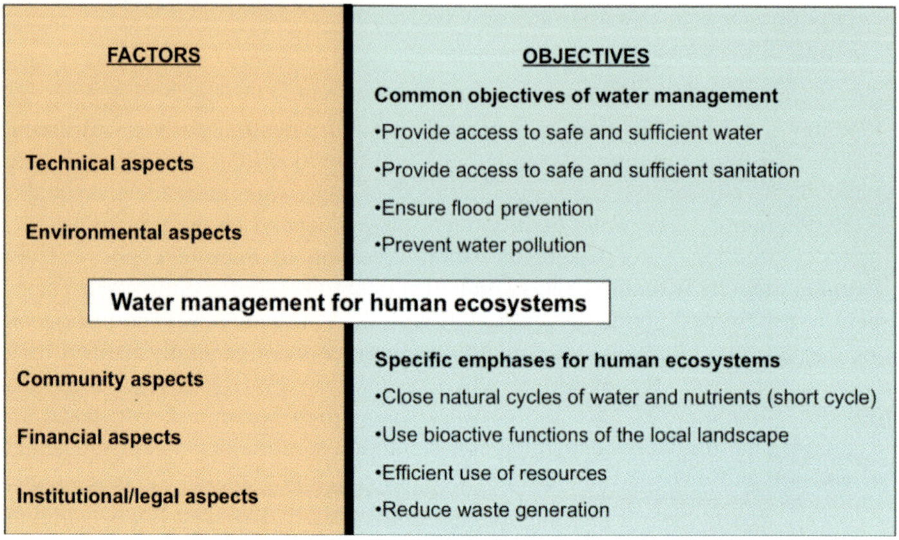

Fig. 3.8 Water management for human ecosystems

For sustainable human ecosystems a different sanitation concept should be applied. The key difference is that wastewater is rather considered as a potential resource than as a waste. Depending on its origin, wastewater contains undesired constituents (e.g. pathogenic microorganisms) but also useful elements (e.g. nutrients). The wastewater can be collected and treated in a way to remove the undesired elements and to be reused for various applications including irrigation.

Ecological sanitation generally is based on separate collection of the different fractions of wastewater. Urine and faeces are usually separated to support reuse. Simple treatment facilities may use local material and can be integrated in local landscape. Ecological sanitation aims at closing the natural cycles of water and nutrients. Ideally the cycles should be as short as possible and therefore water and sanitation systems in human ecosystems are often decentralized. Water safety regarding water borne diseases has to be a principle for design and operation. Sanitation concepts should prevent water pollution. In addition, protection areas for (drinking) water production should preserve the current or future use of resources. Finally, flood prevention should be integrated to existing landscape as much as possible (Fig. 3.8).

Different solutions can be adopted to reach the above-mentioned water management objectives. These solutions are usually based on the use of technologies, implementation of management practices and close cooperation with the community.

However, the first step is to define the problems and development goals including stakeholders and consumers. Environmental, community, institutional, legal, technical and financial aspects must be considered here for better identification and formulation of expectations and constraints. In the field of water management one should keep in mind that the solutions have to be adapted to the local demand and context.

It is obvious, that water is a natural resource essential to life and development of livelihood and society. Still one should always be aware of the important work necessary to implement, operate and maintain water services. These depend not only on technical/physical infrastructures but strongly on software like human, institutional and organisational resources.

The sustainable operation of water services e.g. depends on financial aspects such as cost recovery. Investment as well as operation and maintenance costs induced should be recovered to safeguard the continuity of a high quality service. Therefore the planners and decision makers must take into account the willingness and ability to pay of the community.

Finally the pertinence of the selected solutions should be verified: Does the solution really address the problem? What are the benefits of the solution? Will the solution be sustainable? Do the community/users accept the solution? Which constraints results from the adoption of the solution? Is the solution economically viable? Etc. The advantages and drawbacks of each solution should be balanced by the stakeholders. Indicators and criteria for performance, reliability, environmental impact and costs can be used to ease the process.

3.5 Disciplinary Approaches for Ecosystem Components

3.5.1 Selection of Specific Functional Areas

If the technical contents are investigated first in cells, partial solutions can be discussed by experts in a less complex way after traditional solutions. Then linking up these solutions must be led in the team step by step to a whole solution. One should choose technical units as usual like energy, water, open space, mobility, social and cultural questions etc (Fig. 3.9).

3.5.2 The Various Fields

3.5.2.1 Energy

The energy consumption is looked as a key factor for a sustainable design of our environment. Leading aim is the extensive use of renewable energy what is to be reached at the moment only with energy saving. A "cell" could be a specific building with its surrounding. For its power demand a broader view will be necessary, in many cases the electric net serving not only the specific building, but also the surrounding buildings and settlement.

For the design and the construction of buildings the means are well known: Energy losses in the heating and cooling case are to minimise by building form, insulation and airing, passive and active use of sun, natural cooling by specific

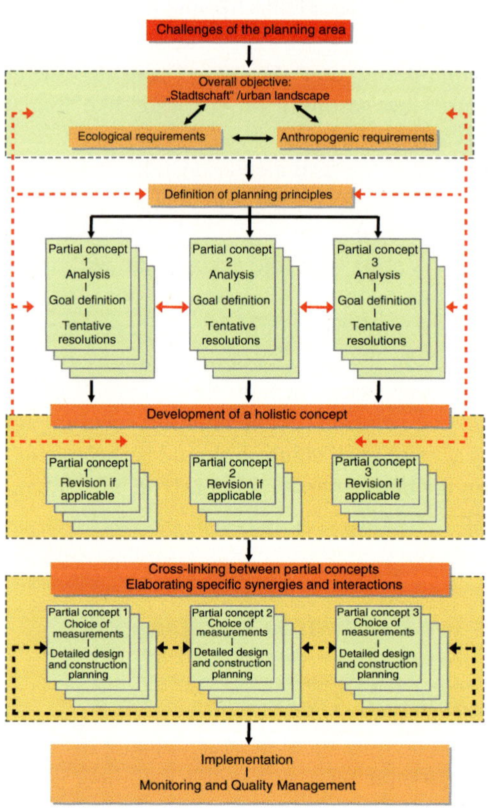

Tools	Results	Stakeholders
Site visit, enquiry, documents and information on planning area	Problem analysis and structure	Planner,client, local authority
Definition „urban (land)scape", environmental data, visions and wishes of users/residents/client	Target analysis, Definition of target characteristics, target planning	Planner, experts, client, township, local authority, environmental associations
Composition of potential planning principles (e.g. water and nutrient cycles)	Priority list for planning principles	Planner, experts
Definition of partial concepts, criteria list assigned to partial concepts	Structuring planning exercise according to partial concepts	Planner, experts
Conceptual design: definition of working planes, assembly of possible targets, collection of all data and information necessary for the analysis	Troubleshooting, problem and goal-oriented analysis, detection of specific conditions, deficits and potentials, investigation of consumption of resources and emissions, composition of specific targets, ideas for solutions, initial decisions for fundamental principles	Planner, experts
Description of feedback loop and consideration process	Discussion of tentative solutions, pre-dimensioning, inducing planning decisions	Planner, experts, client, township, local authority, environmental associations
Key figures, description of scenario analysis	Synchronised partial concepts, compilation of contents of each partial concept, pre-dimensioning, scenario analysis, comparison of different versions if applicable, design	Planner, experts
Assembly of measures and classification in categories, description of characteristics of holistic concepts, example calculation	Discussion of measures, tuned set of measures, realistic integrative and holistic concept, corresponding detailed design and construction planning, calculations, dimensioning	Planner, experts, client, township, local authority, environmental associations
Manuals, operation and maintenance instructions, interviews of users regarding acceptance	Project realisation, construction, target validation	Operator, contractor, quality manager, residents if applicable

Fig. 3.9 General procedure for integrating specific areas into the holistic concept (**Glücklich** 2010)

ventilation, plants, day and night balance, earth register among other things. The minimisation of the energy consumption in the building is not sufficient to exhaust the field of Energy. He is determined by many other parameters.

The climate change is mostly seen in direct connection with the energy consumption and the CO_2 amount. Nevertheless, an important influence on the global warming effects is the bioactivity over the whole word. It leads moreover to the storage of the CO_2. The bioactive green space is important for the microclimate in the building and moreover in the whole ecological system. The grey energy of the materials—mostly in the building area is overrated. The material flows are to be looked and not only to carry out a CO_2 consideration.

The kind of the mobility (transport) has big influence on the energy consumption. Already with the relatively easy field "Energy" the complicated interlinking of the connections becomes visible which must be processed with the development of ecological whole concepts. With intelligent local energy concepts expenses come down, the supply is stabilised and with it the base quality can be improved.

3.5.2.2 Materials

Essential leading aim is the sustainable close of the material cycles as well as the sustainable management of the limited resources—namely in the short-term consideration of the everyday rhythms of life (nutrients, consumer goods) as well as in the long-term consideration as for example the construction of buildings and the infrastructure.

In small cells like buildings should be used corresponding materials, nevertheless, the sensible reasonable choice of the materials can be valued only in the bigger cycles. Because we handle at present wastefully with the resources, is the reduction of the material use and the use materials which can work in cycles has a preference also without long investigations. The CO_2 optimisation is problematic from the date basis and the circulatory effect here. With an intelligent cycling of materials the resources like energy and natural materials is saved and with it the base quality is improved.

3.5.2.3 Water and Nutrient Cycles

Essential leading aim is a sustainable local water cycle which is to realise local and in large ecosystems. With a working water cycle many aims can be reached: High bioactivity and with it climate balance (cooling effect in the micro and macro area), water retention and with it improvement of the ground water situation, use of nutrients, nutrient circulation with high resource use, improved land-economic production on sustainable base etc.

The water cycle can be built up within buildings (small cell), e.g., by rainwater use, water savings technologies and water recycling, urine separation. Buildings become supplier of water and nutrients. Through this the End-of-Pipe-Strategy is changed, the infrastructure is cheaper. Water outside of the buildings should be led in the bioactive space as near as possible to the plants. The function of an intelligent water cycle is important for the survival of the people. The worldwide most important problem is to be approached.

3.5.2.4 Vegetation and Agriculture

This field is closely connected with the water household—as above explained. However, it is also influenced very strongly by the mobility, because traffic routes cut the vegetation and complicate her interlinking. The cells which allow a stable vegetation in themselves namely of very different kind has to be arranged. A part of the cells serves for production of different kind like garden products, flowers and other useful plants. The kind of cultivation will be different to the common agriculture, because the land is used multifunctionally. The vegetation is an absolutely essential part of the Stadtschaft.

3.5.2.5 Mobility

The motorised traffic with its multi-layered environmental charges as well as bicycles and pedestrians take a large part of the cultural space and come to conflict in particular with the water cycle and the vegetation. The traffic ways mostly have no multiple function in the linked up ecological system as for example water. They need from 30% to 40% of the urban space—in particular by the individual cars in the towns.

Leading aim must be to form out the transport function of the traffic routes optimally with minimum influence on the ecological system and to improve at the same time the quality of life substantially. The traffic routes must be looked as a kind of bridges which are going through the ecological system without affecting it substantially—like the walkway trough a moor.

Simple means: Town planning with the short ways, separation of motorised traffic from pedestrians, bicycles, small electrical transporters, carfree park-like cells in the town cells to live and work with car park machines and mobility knots on the edge, high-capacity street system free of traffic jam and a public transport system with an automatical good transport system. In this field a lot of developing to be performed.

3.5.2.6 Socioculture

The economic, social and cultural circumstances form the man-made landscape. The social infrastructure like school, hospital and sport ground must be installed after sustainability.

The cultural aspects are important, because we build up our identity and our life on it. The requirements of the environment are irrefutable. They are not yet fixed in our world view and our cultural norms. Only if they have joined with our norms and the social bases, we are enabled for consistent action. Therefore, our life culture must change without losing the roots—a difficult process.

There already are in particular in the town planning attempts as civil participation, media gates, master courses, workshops, environmental centres etc. The solution is still difficult, because the aims are not clear and our moral norms ere not developed in this field. Civil questionings show the dilemma: The citizens don't know how to live sustainable. They have not enough experience and education. A long process of decision making is necessary. More complex solutions are necessary instead of single activities.

3.5.2.7 Economy

Ecological measures are often not realizable apparently because of costs, so they say. The basis of our life is the nature, so the statement is basically wrong, paper money cannot be calculated against the natural requirements.

Economic decisions are mostly made for short time. However, sustainability is a long time process. Besides, manufacturing, laws and way of life are not conformed to nature, successful short time activities can lead to damages. Nevertheless, in many, almost to most cases these are absent knowledges and habits which hinder the economical and ecological development. Ecology is the frame and not a part of economy.

3.6 The Ecological Masterplan

If the cells are harmonised in groups and at last in a consistent whole solution, the result must be fixed for ecological design and building: A master plan has to be developed. This is a planning process. Moreover serves the master plan to fix and explain results. It is the frame for an architectural development. The results are condensed in the master plan. In particular the spatial modelling of the ecological system is held on in it, buildings, open space, traffic ways and other facilities like water reservoirs, green classroom, gardening and farming, particularly protected areas are part the master plan. He can also contain an accompanying text.

The master plan must be clear for the partners and be well readable. Symbols illustrate functions. The master plan can be changed only by people who have created or have at least good knowledge on this field. They should be able to use the tools to create integrated ecological over all concepts.

Chapter 4
Design, Implementation and Operation of Human Ecosystems: The Example of Valley View University (VVU)

4.1 Study Area

4.1.1 Geography and Landscape Context

4.1.1.1 Location and Topography

VVU is located in the peri-urban outskirts of Accra, the Capital of the West African Republic of Ghana (5.79°N, 0.12°W). The core city, the Accra Metropolitan District, has a population of 1.8 million, while the Greater Accra region is home to 2.9 million. The Accra region is currently one of the fastest growing urban agglomerations in Africa with a growth rate of more than 3% (United Nations 2008).

The city is located in the Accra plains (Fig. 4.1). This lowland area in SE Ghana covers around 2,800 km^2 bordered by the Gulf of Guinea to the south, the Akwapim range in the northwest and the Volta River to the east. Here, the Plains merge with the savannah corridor of the so called Dahomey Gap, which stretches further from SE-Togo to W-Nigeria. The topography is mostly flat to gently undulating.

Valley View University is situated around 30 km northeast of downtown Accra just outside the borders of the Accra Metropolitan District, at the radial road that leads to the town of Dodowa, which is 15 km to the north. The altitude is around 80 m.a.s.l. However, only a few kilometers to the west, the Akwapim range rises steeply to an elevation of 400 m.

4.1.1.2 Social Environment

The VVU campus, 120 ha in size, is surrounded by old villages and new residential areas. It has borders with the communities of Oyibi and Otinibi.

The old villages are rural in nature, often lacking infrastructure in terms of sufficient electricity and water supply. Dusty roads in poor conditions connect the

G. Geller and D. Glücklich (eds.), *Sustainable Rural and Urban Ecosystems: Design, Implementation and Operation*, DOI 10.1007/978-3-642-28261-4_4,

Fig. 4.1 View into rural areas of the Accra plains from the Akwapim mountains

villages to each other and to the main road. The new residential areas, in contrast, represent a rather new type of urban development for African cities with a well-developed infrastructure and wealthy to middle-class residents (Yeboah 2000), which are very similar to western suburban areas. However, as for all of Accra, there is a number of environmental problems associated with the general poor sanitation and waste management standards in both types of neighborhoods in the fast growing peri-urban areas of Accra (Yankson and Gough 1999; Owusua 2010).

4.1.1.3 Climate

The climate in the Plains is known as the Ghana dry zone or the Accra-Togo coastal climate. The Accra plains and the Dahomey Gap represent a hiatus, a discontinuity in the West African (Guinean) forest zone with unique climatic conditions. Accra receives only 750 mm rainfall, which is not enough to support the rainforests typical of the neighboring Upper and Lower Guinea forest zones with rainfalls between 1,300 and 5,000 mm. To illustrate the steep climatic and ecological gradient: Aburi Botanical Garden, on the top of the Akwapim range and only 7 km away from VVU, receives 1,200 mm; VVU, in contrast, gets even less rain (700 mm) than Accra due to its location in the rain shadow of the Akwapim range.

The precipitation is distributed among the two rainy seasons (May–Mid-July, Mid-August–October) and falls often in short but intensive terms, which causes frequently local flooding.

During rainy season, the monthly precipitation is between 50 and 150 mm. The period between December and March is especially dry, with monthly total rainfall ranging from 0 to 50 mm. Despite this seasonality, the amount of rainfall is highly variable from year to year. Sometimes there is hardly any rain during rainy season, while in other years the precipitation is well above the average, even in dry season. Hence, land use management, farming and water management in the area have to cope with this unpredictability.

As to be expected for a tropical country, temperature shows little annually and daily variation, the mean annual temperature being 26.5°C. Monthly averages range between 24.5 (August) and 28°C (March), the average temperature during the day is 30°C. Humidity is high in general (65–95%), but is lower during the warmer months, particularly in January with its dry northeast "harmattan" winds.

4.1.1.4 Soils and Agriculture

The Accra plains are an ancient cultural landscape, inhabited and formed by humans for 6,000 years. The dominating soils in the region are savannah ochrosols and regosolic groundwater laterites over acidic and basic Gneiss and Shists rocks (Brammer 1962). The reddish, brown and yellow sandy to loamy soils are generally shallow and prone to erosion. Organic matter content and nutrient status is low. However, plant growth is even more impeded by the poor physical properties of the droughty soils. Crops grown in the area without irrigation have thus to be drought resistant. A grassland savannah, traditionally grazed by cattle and regularly burned, is hence still the predominating vegetation and land use type in the Accra plains (Figs. 4.2 and 4.3).

Despite Accra with its large market for perishable food crops, the plains have not become an important food-producing area. However, in peri-urban areas, staple crops, mostly cassava and corn, are traditionally grown on numerous small fields in permanent cropping. Basic backyard farming is important for the urban poor. Commercial cultivation of vegetables provides some income for small enterprises. In recent years, there is an increase in commercial mango plantations in the more rural areas.

4.1.1.5 Natural Vegetation and Land Use Conflicts

Savannah grasslands and thickets were the most common vegetation types in this area prior to recent urbanization. However, the savannah is usually coined as "subclimax vegetation", a "derived savannah" that does not mature into a forest due to anthropogenic influence. The natural vegetation of the Accra plains is a dry semi-deciduous forest of various types, the most unique type being the "south-east outlier type" described by Hall and Swaine (1981) from the nearby Shai Hills reserve. This type is forest at its driest extreme in Ghana. The vertical structure features a closed canopy composed of mainly *Millettia thonningii* and *Diospyros sp.* that reaches only 15 m, while *Dryopetes parviflora, Vepris heterophylla* and numerous savannah thicket and liana species form a 10 m understorey. Ground cover is sparse. Scattered emergents tower up to 50 m above the canopy (*Ceiba*

Fig. 4.2 Grazed savannah with Neem trees

Fig. 4.3 Overgrazed savannah during dry season

pentandra, among others). The remaining stands of this forest type are today restricted to scattered, very small reserves (Figs. 4.4 and 4.5).

The original dry forest cover of the Accra plains has been turned into savanna by continuing human impact over the centuries. Regular burning, grazing, firewood cutting, charcoaling and rotation farming have maintained a savannah mosaic with varying proportions of grass and thicket cover. Regular disturbance prevents the development of forest. The thickets of the Accra plains are a rather unique mixture of savannah and forest species in the context of African tropical vegetation (Lane 1962; Jenik and Hall 1976; Hall and Swaine 1981) and, just as for the forest, stands of these semi-natural vegetation types become increasingly rare, due to the rapid

Fig. 4.4 Natural dry forest vegetation in the Accra plains (Hall and Swaine 1976, modified)

Fig. 4.5 Natural dry forest at the Shai Hills reserve

urbanization that leaves not much space for farmland and natural/semi-natural habitats.

Many farmland and remaining savannah areas are already surveyed and doomed to development. Even the existing reserves in the plains are endangered by encroachment. This puts pressure on the remaining land and the traditional land uses. Overgrazing by cattle herds, too frequent fires caused by the pastoralists, as well as firewood cutting and charcoaling by dwellers threaten the ever decreasing areas of semi-natural and natural habitats.

4.1.2 Socio-Cultural Background

Valley View University is a Christian university, that means, part of that Christian socio-cultural background. It is also a Ghanaian university, that means part of the Ghanaian and African cultural framework. And it is also like the general Ghanaian and African (and worldwide) system in the middle of a fast change, where many of the traditional values diminish and new values yet do have to evolve.

4.1.3 Administrative Background

The administrative system in Ghana seems to give more liberty concerning permits, for building for example. After introduction of a general masterplan for the campus and having the general ok of the respective authorities for the outlined measures the further implementation can be put in place quite easily. Instruments of city and regional planning and the respective masterplans are not existing in the same extent and not handled as strict as in Germany. This makes it much more easy to implement an ecological development with a lot of maybe new approaches and technologies and in a much shorter period of time compared to Germany. Some of the measures might not have been possible in Germany in the necessary time span at all.

4.2 Case Study VVU

4.2.1 The University and Its Campus: General Background

Valley View University (VVU) was founded in 1979 by the Seventh-day Adventist church as a school of higher education. In 1997 VVU became the first private university to be accredited by the National Accreditation Board of Ghana. The number of students at that time added up to about 90.

The major programs 2010 are accounting, religious studies, theology, computer science and nursing. Meanwhile, around 3,500 persons are working and studying on campus. About 37% of students are female and 5% are foreigners from West African countries. Circa 20% of the students live on campus in student hostels, the other students live off-campus in the nearby communities.

4.2.2 The Ecological Development of VVU

Human beings everywhere do rely on natural resources like water, soil, air and vegetation, therefore the necessity of a sustainable, conservation-minded development is obvious, particularly in tropical regions like Ghana, where the natural environment is exceedingly vulnerable. Thus it would be very supportive, to have a showcase of such a truly holistic ecological development.

The first visit of German partners of the Ecological Engineering Society IÖV at VVU 2001 made clear, that for this university and its campus being still in its starting phase an ecological development would be a big chance. Both for the Ghanaian and German partners.

From the Ghanaian perspective it would give the university an unique and outstanding position as maybe the first real eco-university in Africa and help to solve a lot of urgent topics like water and energy supply. For the German partners it would allow to test and improve integrated and sustainable concepts under real-world circumstances.

Because of the common interest on both sides for an ecological development of VVU the decision was easy to start a close cooperation and look for the financing. Even at that time first works for an ecological masterplan could be started at the Bauhaus-University Weimar. The proposals and sketches among others served as the base of intensive discussions with the responsible members of VVU and later of the realized final masterplan.

The application for a Research & Development (R&D) project in a tender for decentralised water solutions of the German Ministry of Education and Research (BMBF) was successful and could start 2003, concentrating on cycles of water and nutrients. It was clear to the project partners however from the start, that the functioning of such a project utterly would rely on its integration into a holistic concept and program. The integral development had to be based on an ecological masterplan for the physical development of the whole campus, on the respective design of the buildings and installations, on a proper design and management of the mass flow, on a comprehensive management of quality and information, that would include education, training of practitioners and measures for gaining acceptance internally and externally.

Not being part of the tender for the ecological cycles, this holistic and complex approach had to be defended by the partners, but was accepted finally and turned out to be indispensible for the success. Due to the complexity and the variety of themes dealt with, a far-reaching, interdisciplinary teamwork turned out to be necessary for the success of the pilot-project. For the implementation simple, affordable, robust, appropriate and up-to-date solutions were given priority over low- or high- tec-solutions. This helped to ensure the necessary acceptance, empowered people to help themselves and combined poverty reduction and environmental protection.

Partners of that program aside of VVU have been the Ecological Engineering Society (IÖV), the Bauhaus University Weimar and the University Hohenheim, and two companies, Berger-Biotechnik and Palutec.

Based on the success of that R&D-project the German Federal Government via the German Ministry of Environment, Nature Conservation and Nuclear Safety (BMU) spent another 1.3 Mio. Euro for investments at VVU in the frame of the Climate Change Initiative. Hereby the ecological development made a big step

Fig. 4.6 Baobab-center for ecological studies at VVU

ahead. It started 2009 at VVU and included large-volume rainwater storage under streets, water treatment, planting of several thousand of climate-adapted trees, conservation of old trees and precious native vegetation and the construction of a climate-friendly building for ecological studies (named Baobab-center).

To utilize the implementations for education was an essential part of the projects from the beginning. The idea was to integrate theoretical insights and practical experiences in new courses of studies and new academic disciplines at Valley View University and at the universities of the German project-partners. The ecological buildings with their ecological installations, the sustainable cycle management and the reuse of nutrients on the farmland, the measures of water storage and management served for purposes of exercise, demonstration, training and as reference for people in- and outside VVU. Since 2002 the projects at VVU have been part of the teachings of Bauhaus-University Weimar and University of Hohenheim.

At the moment the establishment of an interdisciplinary master course in ecological engineering and development still is in preparation. The Baobab-center is home of the educational activities and at its inauguration was the place of the International workshop on climate-friendly settlement-ecosystems 2010 (Fig. 4.6).

4.2.3 VVU as a Human Ecosystem

VVU is a representative human ecosystem with all the natural and technical-cultural parts: the campus is a large ecotop, still with a lot of natural elements. It houses producers like trees and greens, consumers mostly as humans, reducers in the soils, but also technical producers as tailoring in the womens center, technical consumers (all the computers, air conditioning, machinery, cooking equipment, television etc.) and technical reducers as dumping-sites, composting toilets, septic tanks etc.

All the fluxes are existing. The natural ones in the form of information as singing of birds or speaking of people, in the form of energy from the sun or wind, in the form of mass flow as rain and erosion of soil. The technical-cultural fluxes show in the form of information as internet, telephone, mail; energy as electricity or fuel; mass flow as wastewater, building material, refuse etc. This all makes up for being a very good showcase of developing a human ecosystem in a sustainable way.

4.2.4 The Ecological Masterplan

The campus of the VVU has many elements which can be taken up in a holistic ecological concept and be developed: On the campus there are many activities beyond the teaching which transport a sustainable living and working. About 15–20% of the students and many employees live on the campus. The numerous ecclesiastical events, excellent music teams and sports teams and student circles like the Green Club make the life attractive, in particular because the African towns in spite of her size have less offers than European ones. A construction department with a block manufacture, small companies like bakery, car repair service and a bank are on the campus. So the developed infrastructure and the workplaces of the Cam-pus make it attractive for the surrounding villages. Besides, students can get some jobs. The everyday ways of the people can be held short, the communication is easy and personal. The campus itself is connected well with Accra and, finally, in particular about the ecclesiastical organisation with the whole world. In connection with teaching research projects can be organised. Therefore: The campus offers good opportunities to live, learn and work in a sustainable way.

The architectural development as a frame for the future life and work should be documented in a master plan which fixes the frame for the ecological development of the campus. A plan already existed for the development of the area with the accompanying buildings (Fig. 4.7): A big oval of approximately 700 m on 250 m should exploit the site with all buildings. The shortages of this plan in ecological regard were evident, repairing is not possible because the fundamentally different kind of thinking. Therefore, a new master plan was developed after the system for the development of ecological concepts over all. Further details see Sect. 4.7.

4.3 Integrated Management of Process, Information and Quality at VVU

4.3.1 VVU and the African Socio-Cultural Framework

The most crucial thing for achieving a sustainable human ecosystem is gaining as complete an understanding of this specific human ecosystem as possible (Fig. 4.8):

When we take the case of VVU, a lot can be learned from traditional African values (which is true for indigenous systems in general). They stem from

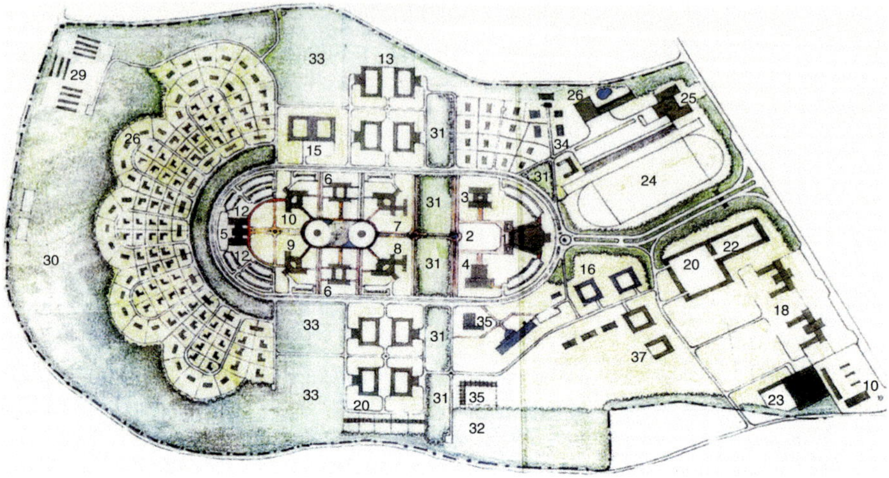

Fig. 4.1 Old master plan from the VVU master plan committee, 2000

Legend

1 Church /Dept. of Theology
2 Ceremonial Ground
3 Administration
4 Central Auditorium
5 Library
6 Central Classrooms
7 College of Arts & Science
8 School of Education
9 School of Business
10 School of Postgraduate Studies
11 Central Plaza
12 Cetral Car Park
13 Femal Students Hostel
14 Male Students Hostel
15 Department of Adult Studies
16 University Practice
17 Security
18 Industrial
19 Fuel /Shopping
20 Commercial Students
21 Maintainance
22 Mini Market
23 Maintanance Yard
24 Multi Purpose Play Field
25 Clinic
26 Gymnasium/Recreational Centre
27 Residence Hall, Dean`s Housing
28 Faculty Housing
29 Poultry Farm
30 Agricultural Farm Land
31 Deforestation (Buffer)
32 Botanical Garden
33 Future Expansion
34 Business Administration Centre
35 Married Students Housing
36 Central Cafeteria
37 Vocational Skills Development

Fig. 4.7 First masterplan without ecological orientation

human-based systems that have adapted very positively into the surrounding natural ecosystems without having too much destructive effects. Among these values are humanity and brotherhood, harmony between communalism and individualism, morality, appreciation of the family, the chief system, arts and aesthetics, etc.

Traditional African culture recognises the dignity and integrity of the human being as a creature of god (this and the following stem from Gyekye 1996). Our common brotherhood is intrinsically linked with our common humanity: There is only one universal family to which all human beings belong. The recognition of all human beings as brothers in light of our common membership of one human species is a lofty ideal that is of great importance to the African people.

The value that traditional African societies place on communalism is expressed in the sharing of a common social life, commitment to the social and common good of the community, appreciation mutual obligations, concern for others, interdependence and solidarity. African ethic urges the avoidance of extreme individualism, which is seen as potentially destructive of human values and of the whole meaning

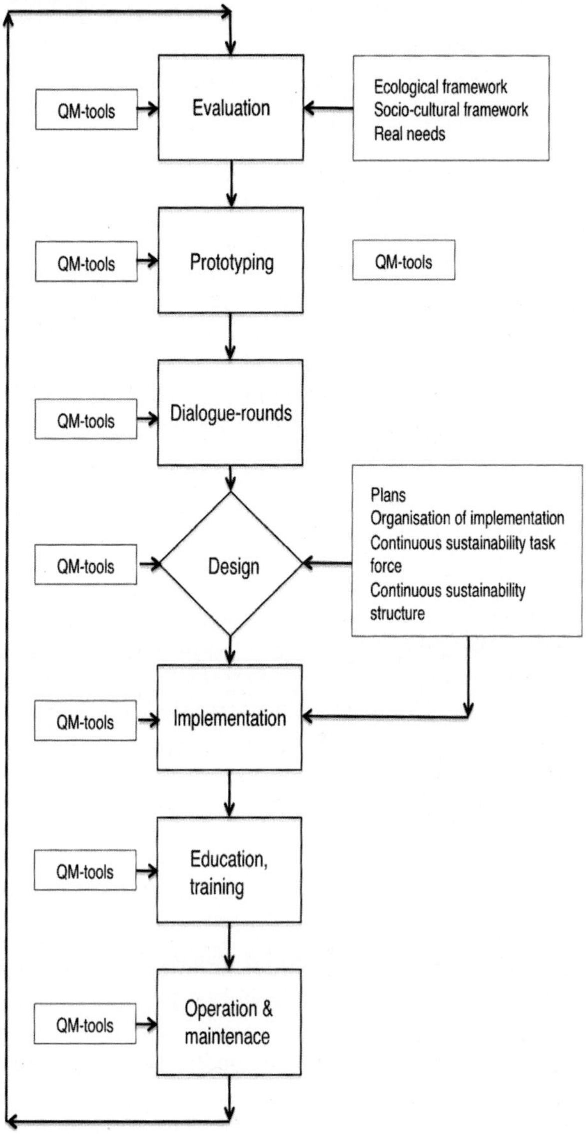

Fig. 4.8 Flow chart of integrated management of process, information and quality

and essence of human society. Attempts are therefore made to balance communalism and individualism so that they can co-exist. Thus, it is clear that African morality is a social morality, not an individual one. The quality and meaning of social life is measured in terms of the extent to which each person responds to the needs of others. The African moral system puts the ethics of responsibility above the ethic of individual rights, even though the latter are also given due recognition. The importance of character as the engine of moral life in practice is stressed.

Special attention is, therefore, given to the cultivation of other-regarding moral virtues, or traits of character.

Family is seen as the basic unit of social life, which includes all of the kinship ties that are essential to a comprehensive social life in an intricate network of social relationships. Values associated with the family include recognising one's responsibility to the nuclear and extended family and respecting parents and elders.

Political thought and practice in the traditional African society places great emphasis on consultation and consensus, for these are central elements of political decision-making. This practice allows the involvement of all people in the political process. The chief, who is the highest political authority, rules with the consent of and in accordance with the will of the people. The political authority of the chief rests on a trusteeship principle that ensures his accountability to the people

In traditional African culture, knowledge is highly valued, and this especially applies to practical or empirical knowledge. The purpose of the exercise of wisdom is to promote and ensure human well-being.

One prominent feature of art (for example, in artistic performances like music and dance) is its participatory character: It is aimed at deepening communal sentiment and awareness. Among the criteria of aesthetic values and judgement are appropriateness and fittingness. Music, dancing and even clothing must be appropriate to the occasion. In works of art, beauty is seen not only in work of art and in the human figure but also in human conduct, in humanity itself, and in a person's character.

The degree to which the VVU ecosystem will fit in with the global ecosystem, and whether or not VVU and the human members of the ecosystem will act in accordance with ecological principles, will depend very much on the extent to which ethics and values at VVU emphasize this ecological core.

4.3.2 Organizing Sustainable Informational Structures at VVU

The structure of organization and flow of information, which also defines responsibilities, acceptance etc., is essential in human ecosystems.

In case of VVU in the last years we have established a special structure dealing with the ecological development. The new established ecological sanitation manager, the cleaning staff, workers and craftsmen, dealing with the sanitary installations, are part of that unit. The physical plant department with its director is responsible for implementing and maintaining the ecological structures and measures. This includes an agricultural section with an trained agricultural director responsible also for the utilizing of the collected waters and nutrients.

To further optimise this structure, a new ecological unit shall be established in the new Baobab-center for ecological studies in the middle of the campus, where the coming Master-study for ecological development shall be housed. The head of the faculty of ecological engineering is intended to be the one, directly responsible for all aspects of ecological development, from ecological constructions and measures to any training and education in this field.

4.3.3 Providing the Necessary Information for a Sustainable VVU

4.3.3.1 Education, Training and Studies at VVU

To gain the necessary awareness and acceptance concerning ecological matters education, training and studies are essential means. Until now, the VVU studies program has mainly covered the areas of business administration, computer science, health services and theology – areas that do not primarily deal with ecology and physical implementation of ecological measures. An International Master studies program in ecological development at VVU is in preparation. All activities in the field of ecological development and sanitation at VVU will be utilized for that and form the basis for a better understanding of ecological principles and necessities. It shall start with an intensive course on ecological basics, ecosystem theory, ecological building, sustainable land use, ecological engineering and more. The MA study will be for BA students from other Ghanaian universities, like Kumasi or and from foreign universities. The partners of the German programs and of co-operating universities like Augsburg, Giessen, Magdeburg, Wiesbaden and Wismar will help establish the program.

In addition to the university studies of ecological development, workers and craftsmen are enabled in the field of ecological building, ecological sanitation, sustainable technologies, etc. by the help of specified training courses. The first steps have been to train personnel working in the VVU constructions unit, ecological cycles program, Climate Change program and that of co-operating firms like Berger-Biotechnik and Palutec.

Meanwhile with the realization of the Baobab-center of ecological studies a location is established where all this activities have a place. Moreover the center in itself is a positive showcase for an integrated ecological solution, producing more energy and water than consuming itself, offering green classrooms and examples of local vegetation and embracing an impressing Baobab-tree to which the center is oriented and its name is coming from. Since VVU is a very positive example of ecological development, it will serve as a demonstration and training unit for all of its members and visitors too.

Aside from that it is essential to enrich all of the other study programs at VVU with the basics of ecology and ecological engineering this generally will be part of all studies and training programs at VVU, including the training programs at the Women's Centre.

4.3.3.2 Acceptance Inside and Outside VVU

Another important aspect and precondition of achieving sustainability of human ecosystems like VVU is getting full acceptance inside and outside.

One example may show that: If the people expected to use the installed new ecological toilet systems and urinals do not use the facilities, the volume of harvested sustainable sanitary materials such as water and nutrients will be small.

The VVU-farmland then will benefit from these measure to only a limited degree, agricultural goods will be few and the profit for VVU poor. The use of the toilets is depending partly on the proper use by the users (sitting toilets in sitting, not squatting mode; leaving toilet paper for the next users etc.), partly on proper maintenance (enough and trained cleaning staff, enough maintenance and operation materials etc.). All this in the first place is a question of information, as communication, as acceptance, as giving responsibilities etc.

A main concern of the ecological development at VVU from the start was to disseminate information to the world beyond the campus on the positive example being set at VVU, including our experience with the project and the lessons learned. One measure was the erection of signposts on university grounds that gives visitors essential information on the various places and installations, with one big sign being placed at the entrance of VVU providing a general overview.

More information is provided on the websites of IÖV and VVU and co-operating institutions, in publications such as this book and in flyers for the various fields of activity, such as ecological buildings, sustainable energy, sustainable agriculture, the ecological master plan, etc.

4.3.4 Integrated QM at VVU: Measures

To ensure the long-term function (sustainability) of an evolving ecosystem like Valley View University, acceptance and taking over of responsibilities is crucial. Various methods have been developed, applied and tested. Among these are Appreciative Inquiry, U-Theory, World-Café, Circle, retreats and workshops, QM-questionaires etc.

Two events in the course of the program may illustrate the application: A joint retreat of all project-partners in 2006 and the International Workshop ClimSys2010 as the conclusive step of the Climate Change program.

4.3.4.1 Retreat in Dodowa 2007

As part of the ecological program at Valley View University this special retreat was jointly organised by Valley View University (VVU) and the Ecological Engineering Society (IÖV). It took place outside VVU at Marina Hotel in the nearby town of Dodowa on February 26th and 27th 2007.

The main aim was to get the university´s full commitment, the essential precondition for the long-term running and success of the ecological program at VVU. The theme of this event was Valley View University as ecological university.

Nearly 40 participants took part in the 2-day retreat. The Ghanaian participants represented the most important constituents of the university: agric workers, members of the physical plant department and of the cleaning staff, students, teachers, pastors, deans responsible for the student hostels and members of the university governing body.

The retreat started with a circle. All participants introduced themselves and expressed their wishes for the coming 2 days. This was followed by world-cafe rounds. The world-cafe brings four to five people together at one table, where they together consider one question together for approximately 20 min and put down their findings. After this, the participants change tables: only one person remains at the table as the host while the others leave for other tables and new persons join the table with the host. The host briefly explains the findings and then the group considers the second question for another 20 min. This is done two to three times for each topic, enabling a quick assembly of ideas and insights and rapidly building familiarity among the participants.

This was combined with the approach of Appreciative Inquiry (AI), which is strictly positive and encouraging and consists of four steps, namely: discovery, dream, design and destiny. In this retreat, the world-cafe rounds were named eco-cafe, reflecting the main theme of the retreat. The questions of the first three table rotations of eco-cafe round one (discovery) centred around the yet existing positive at the Eco-University VVU. The second step, "dream", created a vision of how the Eco-University VVU could look if there were no constraints (financial, personnel, etc.). The third step was about the possible "design" of the Eco-University VVU, which took into account the existing positive core and the visions for the future. A lot of very interesting suggestions for the different areas of implementation were proposed. After the hosts of the tables presented these in the general assembly room, all the findings were summarised in a mind map, where the most important things could be seen at one glance.

Here each participant could assign two points to each of the noticed topics and thus the priorities of the group could be clearly seen. In the last step of the process, the measures, responsibilities, performance indicators, time limits and supporting organisations were worked out.

The work during the retreat was enriched by an excursion to the showcases of the existing ecological cycles at the VVU campus. These included the sanitary block, urine tanks, rain water and grey water storage tanks, the impressive trials in the agricultural area, the biogas plant and the biogas cooking area in the cafeteria kitchen, and the solar plant for the production of heated water for the kitchen. The excursion was ended around the largest tree on campus as a reminder of the non-human beings in the ecosystem.

The retreat came to a close in a final circle under the big bamboo tree in the Marina hotel garden. Here each of the participants gave a resume concerning their original expectations, their impressions and a personal outlook for their future participation in the ecological development of VVU and their respective activities. All participants took the impression home that this retreat was a big step forward not only to a real commitment for the ecological development of Valley View University, but also for a more effective cooperation of all stakeholders at VVU in general.

The joint ecological program that is still in effect today demonstrates the positive effect of these kinds of activities in the field of holistic information and quality management.

4.3.4.2 ClimSys2010: International Workshop on Climate Friendly Ecosystems

As part of the strategic approach to ensure the long-term function of the measures implemented at VVU, and especially those in the frame of the Climate Change program, an International Workshop was organised as the final step. The presentation of the ecological measures at VVU should broadcast the realised existing examples, thus helping to multiply it. It should also enhance their acceptance inside and outside VVU and help to collect insights and recommendations from among the participants for the design, implementation and operation of sustainable ecosystems in general. To enable the necessary interaction and open atmosphere, it was intended as a workshop with only a few presentations.

The full title was: International Workshop on Climate Friendly Ecosystems. The main topic was the question: Are there successful and holistic climate friendly solutions which can serve as examples for others? These examples should illustrate a real integration of ecology, economy and sociology, that was intended from the start and had been realised at least to some extent. Furthermore, the solution should encompass holistic information management that includes deep mutual understanding of the various traditions, life styles and socio-cultural backgrounds. This integration into a new system should be in harmony with the surrounding ecosystems and with the global Gaia ecosystem.

On the basis of the VVU example and others, we knew the answer is yes: There are successful holistic solutions for human settlements, for water treatment and cycles, for harmonious cooperation, and for real dialogue. Therefore, the task was to look for more examples, sharing our know-how and collective wisdom about sustainable ecosystems in dialogue, and creating visions and ways to implement more successful sustainable solutions.

The participants came from different fields, thus representing the various stakeholders in the processes of establishing sustainable ecosystems. The following fields were covered, among others: architecture, civil services, education, renewable resources, research and development, social science, watershed management and operation and maintenance. Participating stakeholders were architects and engineers, pastors, politicians, researches, teachers and students, workers, people from authorities and NGOs. To ensure an open atmosphere of enthusiasm and a willingness to share ideas, insights, and good and bad experiences, the positive approach and methods of the 2007 retreat and of the International Symposium ecosys09 at VVU in spring 2009 were used again and further improved. Thus, the approaches of Appreciative Inquiry and world-café, and methods like world-cafe rounds, mindmaps, excursions to existing positive examples, were applied once again, this time adapted to the environment of a meeting that was not primarily

Fig. 4.9 Opening ceremony of the ClimSys2010 Workshop at the Baobab-center

internal and for the international participants, coming from a broad field of activities and experiences.

The workshop took place at VVU, from Tuesday 23rd to Thursday the 25th of February, 2010. In the opening ceremony, the Ghanaian Ministers for Family and Gender and for Environment and Technology and the representative of the German Embassy all pointed to the necessity of sustainable solutions and showcases like VVU and expressed their wish to multiply this kind of solutions, for example as eco-villages in Ghana. At the end of this ceremony the Baobab-center for ecological studies as part of the measures financed in the Frame of the Climate Change Initiative of the Federal Government of Germany was handed over from the German to the Ghanaian side (Figs. 4.9 and 4.10).

In the afternoon special presentations were given to the students and staff of VVU, utilizing the Arche Noah picture. After that all the realized examples have been subject of an excursion, where all the participants were given an introduction by the respective project-partners on site. The last action was a tree-planting event, executed by the students themselves.

The following day the workshop as such for a limited number of experienced participants started with several presentations of the realized measures, namely of the Baobab-center for ecological studies, of the tree-planting and conservation of vegetation, the long-term rainwater storages below roads and the water treatment and sachet water production. The following workshop program consisted of deep exchange between the participants and excursions to the examples at VVU, in detail explained by the partners and discussed with the experts. An additional excursion to an African Orphan Aid village showed the realization of ecological solutions under differing circumstances of a sort of eco-village.

During much of the workshop, participants exchanged experiences, ideas and know-how, for example in world-cafe rounds. The Appreciative Inquiry approach was applied in its four steps: joint discovery of existing positive solutions; joint

Fig. 4.10 Official handing over of the Baobab-center

dreaming the vision of harmonious climate-friendly ecosystems; joint designing of such ecosystems with evaluation of the key factors for success and forth, the joint defining of the first real steps. In all of the program, music and Ghanaian cultural performances played an important role.

The last day served to summarise the findings and included a colourful Ghanaian Closing Ceremony. As part of the quality management effort to learn and further improve information management, questionnaires were used at the start and at the end of the symposium.

Some improvements concerning the organization of such meetings, learned of the ecosys09-symposium, could been executed: the organizing committee consisted only of six members, committed and of their own free will participating. Money was made available directly by the climate-change program and the members additional meetings and work in their free time being balanced by some small chargeback.

During the After-workshop tour, the remaining participants were able to broaden their views, deepen their personal relationships and debate even more deeply about future joint projects. The excursion should go to the Songhai-example in Benin. This excursion was special in various regards: Albeit a relatively small distance in kilometers, the time for getting visa and pass only one border made it impossible to reach the destination in the scheduled time. This was influenced additionally by the dangers of passing the road towards Nigeria, where only recently severe events had happened. Instead of clinging to disappointment because of this failing of reaching the destination the participants showed their best sides, utilized the additional time of exchange and made the excursion a positive collective experience, which also deepened the relations between the members.

4.3.5 Main Findings

The topics of integrated information and quality management in its deepest and broadest sense and long-term view are decisive for the success and sustainability of development projects generally. Thus they must be included from the start and throughout the whole project. The approach and tools applied at VVU have been useful. The issue and its application in the field of ecological development being largely new makes it necessary to see it as a continuous learning process. Here the best approach and the best fitting tools always have to be found out in the course of the project in intensive dialogue with all of the stakeholders and adapted to the specific case everytime anew.

4.4 Land Management: Vegetation

4.4.1 Nature Conservation and Biodiversity Management

4.4.1.1 Land Use on VVU Campus

The Valley View College, which was to become Valley View University in 1997, moved from a former urban location to its current Oyibi site in 1989. The new 105 ha (300 acres) property was bought from Ga chiefs, families and private landowners. According to local traditions in land tenure, it was leased out for 99 years (Owusu-Mensa 2009). The land was then composed of savannah thickets and farmland, including fallow land. This pattern of land cover was typical for the rural areas surrounding Accra (Sect. 4.1.1). Some parts of the property were at that time also still covered with tall Kyen-Kyen trees (Bark cloth tree, *Antiaris toxicaria* Lesh.), which are characteristic for the native dry-forests of the Accra plains. It is said that even after the university had started, monkeys fed on these trees until they were felled in the 1990s. However, at the beginning of the eco-projects in 2003, large areas of the property had not been managed at all by the university, except for a few buildings, some roads and agricultural land. The "bushland" was regarded by many students as hostile "wilderness", whereas for rural dwellers the freely accessible land was an important source of firewood and charcoal, which was also regularly burned, and grazed by cattle.

The first steps to account for the ecological functions and services provided by the vegetation in the eco-development of VVU (3.3.1) was the design of an extended, unspecified "green belt" in the Ecological Masterplan (Fig. 4.48) and the planting of various large tree orchards in the agricultural projects starting in 2003. The campus witnessed a rapid development of the university, and by 2008, 27% of the total area were already covered with buildings and associated green space. Another 21% were used for agriculture, 29% were awaiting future development and approximately 12 ha were reserved for the green belt.

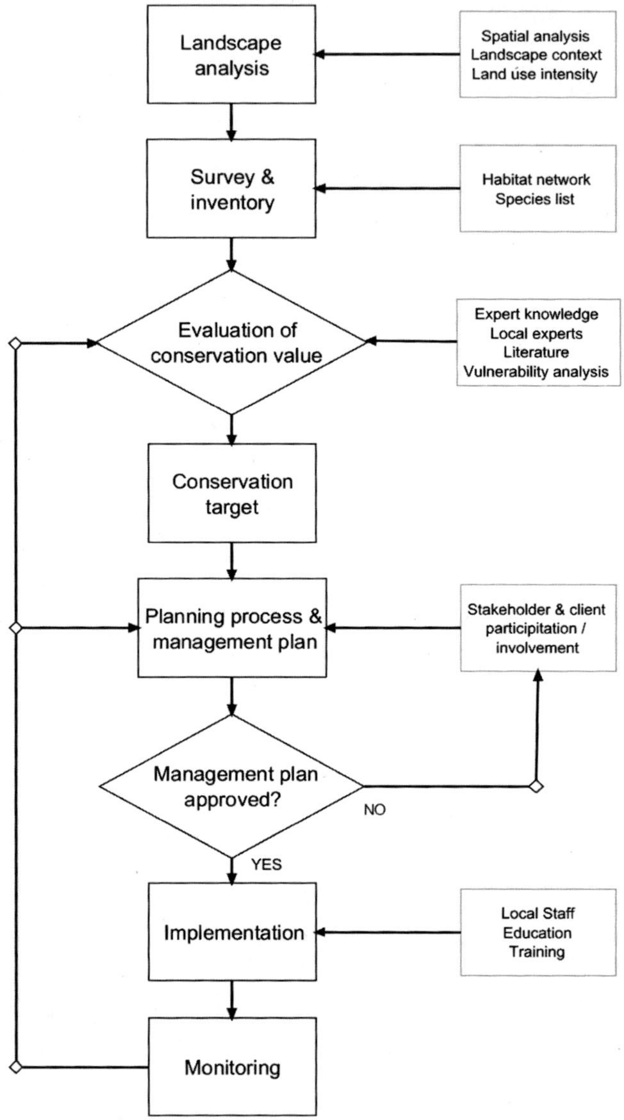

Fig. 4.11 How to integrate nature conservation and biodiversity human ecosystem management

4.4.1.2 Landscape Context

Due to the increasing urban sprawl with ever expanding residential areas and an associated overexploitation of natural resources (frequent burning, overgrazing), only small portions of savannah bushland and natural vegetation are left in the former rural Accra region (see details in Sect. 4.1.1). Larger portions of relatively undisturbed savannah thickets and dry forest are only preserved at the nearby

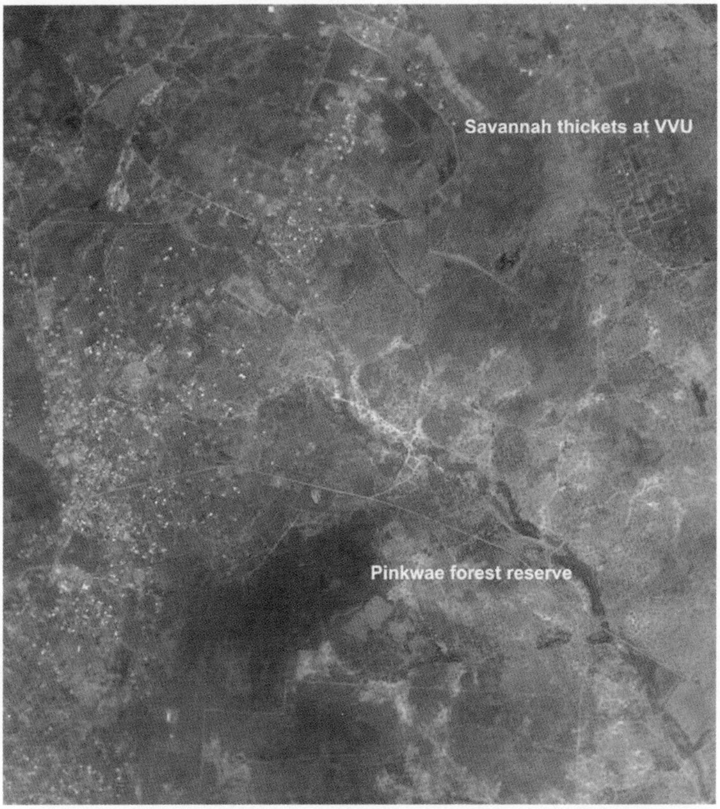

Fig. 4.12 VVU and Pinkwae forest reserve represent last fragments of semi-natural thickets and forest vegetation in the area (*dark color*; Google Earth)

Pinkwae Sacred Grove (Fig. 4.12) and in the Shai Hills Resource Reserve (Fig. 4.5). Considering this development, the remaining areas of shrubland on VVU are most obviously very valuable in terms of biodiversity conservation at the regional scale. However, despite this situation, a true evaluation of the conservation value of the bushland was not part of the initial phase of the project, due to project-related circumstances.

4.4.1.3 Evaluation of Conservation Value

A detailed floristic inventory of the savannah bushland at VVU was carried out in 2009 and with the support of local botanists (Figs. 4.13–4.16). A surprisingly large number of 117 plant species were found in the thickets. Ninety-five of these species had to be considered as species native to the Accra plains and typical of the natural wooded savannah. Among them were 26 shrub species, 23 climbers (lianas) and 26 tree species. Out of these, 27 species are regarded as "endangered medicinal plants"

Fig. 4.13 *Dichapetalum madagascariense*

Fig. 4.14 *Millettia thonningii*

on a national scale. Most of them are rare due to habitat destruction and overexploitation for medicinal purposes (Abbiw 1990, no year). From the survey results, the intent evolved to protect large portions of the threatened species-rich bushland on VVU property.

4.4.1.4 Vulnerability Analysis

Many of the recorded tree species in the bushland were only found in small numbers and/or in juvenile or coppiced individuals. Full-grown trees did not occur at all for

Fig. 4.15 *Clerodendron capitatum*

Fig. 4.16 *Scadoxus multiflorus*. All are attractive plants of the savannah bushland

most species. This is a consequence of the unauthorized cattle grazing, firewood chopping and charcoaling by rural dwellers and the frequent fires set by herdsmen and hunters (Sect. 4.1.1) (Fig. 4.17). Another threat to the species-rich thickets is the continuing spread of introduced, dominant species, namely the Neem (*Azadirachta indica*), (Fig. 4.18) and Leucaena (*Leucaena leucocephala*) trees, as well as the two noxious shrubby climbers Siam Weed *(Chromolaena odorata)* and Spanish Flag (*Lantana camara*). Areas mainly occupied by these weeds are however located in those areas destined for future development. The sites harboring the largest portions of species-rich bushland, in contrast, were fortunately located between the actual (planned) campus and the agricultural fields.

In the initial master plan (Fig. 4.48), these areas were already destined as part of the green belt. Therefore, only slight modifications of the initial planning were necessary to permit the preservation of relevant areas of species-rich thickets, without compromising the development of the university (Fig. 4.19).

Fig. 4.17 Threats to the species-rich bushland: unauthorized burning

Fig. 4.18 Threats to the species-rich bushland: invasive species (Neem overgrows native shrubs)

Fig. 4.19 Location of preservation areas (Google Earth)

4.4.1.5 Conservation Target and Management Plan

The idea to protect larger portions of bushland resulted in the delimitation of 18 ha preservation areas (Fig. 4.19). The conservation target is to allow the succession from the current thickets to a natural dry forest with grown trees (Sect. 4.1.1; Fig. 4.4). To achieve this goal, it is of outmost importance to manage the external threats (wood gathering, fire, grazing) and limit these disturbances to an absolute minimum. Further, it is important to halt the spread of invasive species in the preservation areas by selective measures.

4.4.1.6 Implementation

In a first step, the actual location and boundaries of the preservation areas were included in the VVU masterplan. Further construction activities will thus respect these areas. Also, the relevance of these areas was extensively communicated to the most important stakeholders.

Fig. 4.20 The hole allows animals to pass through

As a first measure, large open spaces that were lacking thicket cover due to previous fires were planted with a large number of tree seedlings ("Tree Islands", in total 2 ha; Sect. 4.4.2).

A further successful step was the implementation of additional security staff patrolling the remote areas of the property. This measure not only protects the preservation areas from people entering the property to hunt or collect firewood, but is also important to safeguard the new tree crop plantations from fires set by herdsmen. Prior to the patrols, four fires were set in the first nine months of 2009 and destroyed parts of the new oil palm plantations. Since the patrols were installed, no more fires were set.

However, to sustainably protect the remote areas of the property, preservation areas and farm land from unauthorized intruders, it will be inevitable to fence the property with a brick wall. Within the project, the first 400 m of a brick wall eventually surrounding the whole property was erected. The wall was equipped with ground level holes (25 × 28 cm) in regular distances to allow small animals to pass through (see Fig. 4.20). Currently, the completion of the wall depends on further funding.

4.4.1.7 Stakeholder Participation

The results of the floristic inventory and their relevance for biodiversity conservation at the regional scale were discussed in an early phase of the project with key stakeholders during a retreat (4.3.4). The location of the preservation areas and the management plan were proposed later and discussed with VVU officials. In general, there was a good understanding, support and acceptance for the conservation goals. Only slight modifications of the initial planning were necessary and not many efforts were needed to declare large parts of the yet undeveloped green belt as preservation areas.

4.4.2 *Tree Planting Project: Green Space at VVU*

4.4.2.1 Background and Aims

The climate in the Accra region with its two rainy seasons is anomalously dry for this latitude (Sect. 4.1.1). Climate models predict an even warmer and drier climate for the region and an increasing unpredictability in rainfall patterns has been observed in the past years already. One way to mitigate the effects of the expected future climate change is to improve the local microclimate by planting trees (e.g. Lyle 1999) (Fig. 4.21).

One of the four pillars of the BMU Climate Change–Project at VVU was hence the "Tree Planting Project". The initial aim of this sub-project was the improvement of the microclimate, in particular the provision of shade and cooling by increasing the number of trees on campus. The planting of a large number of trees was also meant to foster the visibility of VVU as a showcase of ecological development both for the students and the surrounding communities. The latter, old traditional villages as well as new residential areas, are often lacking a substantial number of trees in private gardens and public spaces, probably due to a rather inappropriate, but wide-spread fear of tree-inhabiting snakes. This lack of trees and green space results in a heavy exposure to the tropical heat and in addition to the often dusty and eroded roads, heavy traffic as well as flawed sanitation and water supply, contributes to the unpleasant environment most of the communities surrounding VVU exhibit. Even before the eco-development started, VVU was known already by guests and students for the lush and recreative environment provided by the extensive green space on campus. The aim of the tree planting programme was to build upon this experience and to further improve the situation of the green space at VVU. Another goal was the integration of the plantings in the nutrient cycling of human waste and the collection of rain water.

Box 4.1 Planting Trees in the Tropics
Availability of Planting Stock

The cultivation and planting of trees in the tropics requires some fundamental deviations from practices applied in colder climates.

In temperate regions, trees are usually cultivated in tree nurseries until they have reached the desired size. Harvesting and replanting is carried out in the wintertime when the trees are inactive or have shed their leaves. In the tropics, we lack a longer period in which the tree's metabolism is slowed down. Hence, the digging and replanting of rooted trees is usually not successful. The utilization of bare-root planting stock is therefore very limited.

This disadvantage, however, is mostly being compensated by the rapid growth of planted seedlings. Planting stocks commonly employed in the tropics are therefore seedlings or graftings cultivated in containers (pots or

(continued)

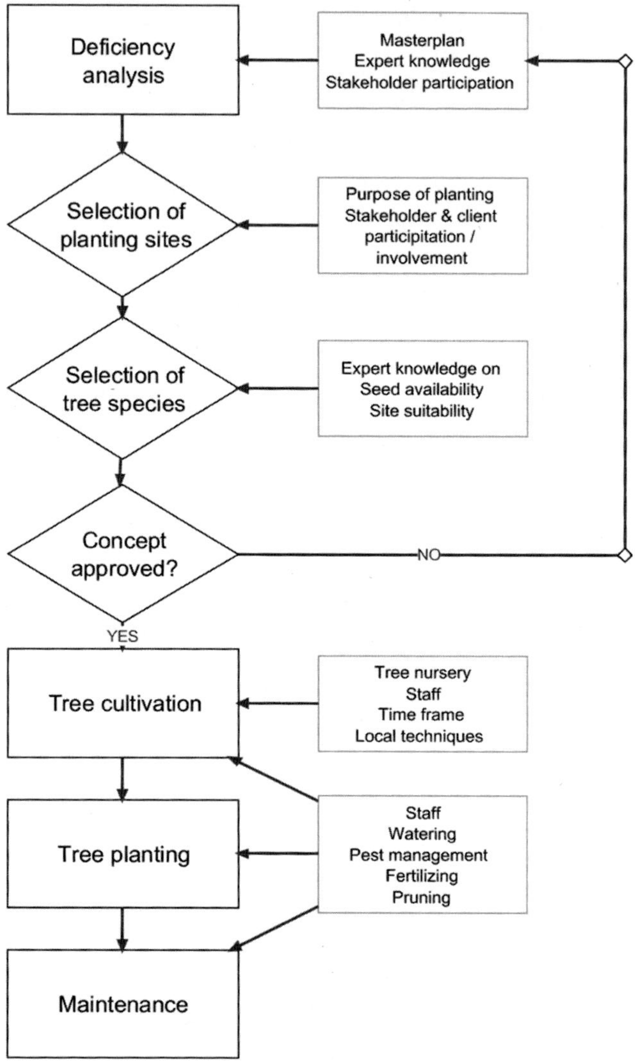

Fig. 4.21 Steps in the tree planting project at VVU

bags). They are usually less than 1 year old with a size of 50–120 cm (cf. Longman 1993–2003; Upton and De Groot 2008). Taller trees can be grown in larger containers but due to the usual lack of costly infrastructure needed to handle grown trees, and the resulting lack of demand, these are not available.

Even more difficult than getting the right size of trees for planting measures is to find the right tree species. In developed countries, trees of many different species and size are kept for sale by a large number of well-sorted commercial nurseries. The number of tree nurseries in developing countries is usually just as low as the variety of available tree species. If nurseries carry tree seedlings at all, they are usually limited to a few fruit trees or popular timber trees like the Indian Teak *(Tectona grandis)*. A broader assortment may be found in Botanical Gardens or Public Forestry Services.

4.4.2.2 Initial Situation and Planning Process

Like in any other design process, the planning of the tree planting project at VVU required a comprehensive, preceding analysis of the initial situation (Fig. 4.21).

The developed campus of the university (approx. 25% of the property) featured already large areas of landscaped green space. These were mainly maintained as turf, carrying however already a small number of older shade trees in some parts. Many roads and paths on the mostly car-free campus lacked shade trees. Some large, degraded open spaces in the bushland were available for restorative plantings and some several hectare fallow land awaited tree crop planting.

In a few first meetings, preliminary suggestions were gathered, ideas were developed and possible options were discussed with the client, namely the President's board and the involved local experts. A resulting first concept was proposed to all relevant stakeholders during an initial retreat and the aims and objectives of the planned measures were clearly communicated. Active participation was encouraged during the retreat and further ideas were collected.

An adjusted and final concept for the planting measures, which contained also some of the ideas gathered at the retreat, was then confirmed by the relevant decision makers and a budget was estimated with the support of the local experts.

Box 4.2 Planting Trees in the Tropics
Grow Your Own Seedlings

The limited availability (see Box 4.1) and the rapid growth of seedlings as well as the comparably simple cultivation techniques suggest to establish an own – permanent or temporary – tree nursery for any larger project in the tropics. There are a number of good manuals and instructions available that deal with the requirements and possible pitfalls in the establishment of a tree nursery (e.g. Upton and De Groot 2008; Harum and Moestrup 2010; see also CL in Sect. 8.4.2). In the case of the VVU Tree Planting Project, it was fortunate to have experienced local experts and gardeners available, who provided valuable expertise on local cultivation practices.

Seedlings are usually cultivated in small polyethylene bags ("poly bags", "rubber bags", "nursing bags"). Depending on the size of the seeds, they are either directly sown into bags, or planted there after their emergence in seedling beds. Common bag sizes are between 300 mL and 1 L. Usually the

(continued)

seedlings get planted at a very small size, only after a few weeks of cultivation, when they start to outgrow the nursing bag. This method however produces many losses.

Hence, more time and effort should be invested into the cultivation of stronger seedlings, particularly when the availability of seeds is limited. To this end, a "rebagging" into bigger bags (4–7 L) and a continuous cultivation for 2–3 months is recommended. This intermediate step was chosen at the VVU project and yielded very good results with little losses after planting.

4.4.2.3 Time Schedule

Due to project-related circumstances, only 1 year was in total available for the planning, the production of the planting stock and the planting. The corollary effect was that there was no time available for a careful selection of tree species and seeds (see Boxes 4.1–4.4). The time pressure made it necessary to already begin with the production of seedlings before the measures were even planned. As a consequence, a number of available species were grown, but proved to be not appropriate, whereas seeds of other suitable species were not available at the time the production had to start.

Fortunately though, a tree nursery already existed on campus and experienced staff was already in employment. The production of seedlings could hence begin in the first month. The planting was carried out from the seventh to the tenth month – depending on the available rain and the readiness of the seedlings for planting (Fig. 4.22).

Box 4.3 Planting Trees in the Tropics
Choose the Right Tree Species

The basic challenge for any tree planting project is the selection of the right tree species. Generally, all chosen tree species have to be adapted to the available water supply and capacity as well as to the soil quality found at the respective planting site. Particularly in arid climates coping with increasing droughts due to climate change, a focus needs to be given to the water demand of the tree species. Here, only drought resistant species should be chosen.

Further, the ornamental value of the selected tree species is important to generate acceptance for tree planting measures in public or private green space. From a landscape design point of view, it is obvious that the chosen tree species should have a growth habit and final size that is in accordance with the intended purpose and the available space.

Depending on the respective purpose of the plantings, other criteria gain additional importance. When the focus is on the restoration of

Fig. 4.22 "Rebagged" seedlings in VVU nursery (Dry Zone Mahogany, *Khaya senegalensis*, an excellent shade tree)

Fig. 4.23 Organic mango plantation at VVU in its fourth year

semi-natural/natural habitats or the general support of biodiversity, exclusively native, i.e. local or indigenous trees should be planted. When the aim is additionally on the production of fuel wood as a renewable energy, a high calorific value of a chosen tree's wood is another important decision support. If the purpose is to generate sustainable income, the economic value and the market situation of the respective tree crops or timber products have to be thoroughly evaluated (Fig. 4.23).

4.4.2.4 Implemented Measures

The tree planting measures carried out within the project fall into five categories, each with a different focus (see Fig. 4.27 for overview):

Fruit Tree Orchards

The majority of trees was planted to complement the existing 13 ha tree crop plantations (mango, cashew), which were already established in an earlier project. Another 6 ha were planted with oranges, mangoes, avocadoes and others crop trees. Particularly the orange plantation (650 trees) is intended as a large scale field trial to demonstrate the practicability to grow oranges in this dry climate. A regular irrigation scheme is absolutely necessary and is taking advantage of the grey water collection in the dormitories. The grey water is transported via pipelines to large tanks installed in the field and applied manually by farm labor. The crops will be delivered to the cafeteria and sold to students and at the local market.

Arboretum / Groves

Early on in the history of VVU, a so-called "Botanical Garden" was established as an collection of different trees (arboretum) planted in wide spacing and accompanied by a well-maintained turf. This established arboretum is already well accepted and treasured for its shade by students, who use it for study groups and recreation in-between classes (Fig. 4.24). Hence, one of the first ideas that came from VVU's side was to enlarge the arboretum by claiming and landscaping the bordering piece of fallow land. This area (2 ha) already contained a number of native savannah trees, which were preserved. Open spaces were additionally planted with an assortment of different shade trees. The ground was cleared and planted with a local, popular lawngrass.

To multiply the shaded areas used for recreation, all areas suitable to support large trees without compromising other interests on campus were detected (2 ha in total). These areas were also planted with different kinds of trees to form eventually shady "groves".

Fig. 4.24 The arboretum at VVU

Box 4.4 Planting Trees in the Tropics
Get the Right Seeds

There are several difficulties involved in starting a successful tree production from seeds. One of them is the limited availability of an sufficiently large number of seeds from the desired tree species. Often, seeds have to be laboriously collected at various sources or purchased from local experts or institutions. Since many seeds are also short-lived and only available for a certain time of the year, it is very important to allow sufficient time for seed acquisition. Great emphasis should be given to the quality of seeds. Only seeds of good quality will produce healthy trees. Poor germination may also be an inherent trait of some species, which has to be dealt with. Experienced staff, sufficient workforce and adequate supervision is in any case very important in this sensitive phase of the project. Longman (2003) gives valuable detail information on this topic.

Perimeter Planting

The entire perimeter (approx. 4 km) was planted with a line of shade trees. The spacing between the trees was 10 m. Once grown, these 400 trees will make VVU a visual landmark in the area and foster the claimed showcase effect as a positive example of sustainable development in the neighborhood (Fig. 4.25).

Fig. 4.25 Perimeter planting with dry zone mahogany (1 year old, *Khaya senegalensis*)

Tree Islands

The open, degraded spaces in-between the existing savannah thickets (see Sect. 4.4.1) were planted with an assortment of mainly native trees (Tree Islands). The conservation aim for these areas is to restore a dry forest to support wildlife and biodiversity. Therefore only rare and native tree species would have been desirable for this measure. Since these were not available in ample numbers at the beginning of the project, compromises had to be made and some exotic, but well adapted tree species were also used for the Tree Islands (Fig. 4.26).

Avenues

Most roads and walkways on the campus lack sufficient shade. Since the traffic-concept for VVU promotes walking and cycling as the main modes of mobility, one of the aims project was to relieve this situation and to provide shade trees along the roads. However, the stakeholder's intentions for the design of the important roads focused strongly on prestigious landscaping. There was a strong desire to plant palm trees, which do not really provide effective shade. However, in order to strengthen the acceptance of the whole project, 300 palm trees were planted along the major roads. Taller avenue trees were planted wherever suitable (Figs. 4.27 and 4.28).

Fig. 4.26 Silk cotton tree-seedling in Tree Island (1.5 years old, *Ceiba pentandra*)

Publicly Effective Supporting Measures

Preservation of giant trees Planting a large number of trees is an effective way to cool the microclimate and to improve the environment in tropical areas. However, even more important than planting new trees is the conservation of old trees in peri-urban areas. At the border of the VVU campus, two giant old trees (Baobab, *Adansonia digitata* and Silk cotton, *Ceiba pentandra*) are basically the sole survivors of a previously larger number of emergent trees that towered above the surrounding, lower dry forest canopy and thicket vegetation of the closer surroundings. In the Accra area, these trees become rarer day-to-day, as the urban sprawl increases. Thus, to make a statement on the necessity to preserve these trees as a natural heritage, measures were taken to permanently save both trees. Located on or just outside the borders of the campus, both sites were bought from the owners, and the authorities were contacted and supported to reroute the road that passes between the property and the Baobab tree (Fig. 4.29).

School trees-project A tree-donation project was launched at VVU to increase the public recognition of the eco-development at the university. Selected schools of poor villages will be provided with tree seedlings (fruit trees, shade trees) to improve the environment for the school children (Fig. 4.30). VVU staff consults the teachers, plants the trees and supervises the maintenance of the trees for the duration of 1 year. Some plantings will be done in a public event involving media coverage.

Fig. 4.27 Overview: tree planting measures at VVU. Approximately 4,000 trees were planted in total

Fig. 4.28 Students get instructions to plant trees during a public event

Fig. 4.29 Baobab tree at VVU border

Fig. 4.30 Oyibi Presbyterian primary school: selected for the tree donation project

4.4.3 Main Findings

An adequate consideration of the ecosystem functions and services provided by vegetation requires a thorough analysis of the landscape context and the semi-natural and natural habitats at the respective project site (4.4.1.3). Due to project-related circumstances both could not be part of the initial projects at VVU, but were accomplished at a later stage. Luckily though, the most valuable parts of savannah bushland were located in areas that were anyway destined for the green belt. Hence, the preservation of 18 ha bushland was not at odds with future development plans at VVU.

One of the main targets of public green space management should be the consideration of biodiversity issues at a project site. This was reflected in the VVU project by the selection of a large number of native tree species and the maintenance of as many grown local trees on campus (e.g. Arboretum) as possible.

Only little more than 1 year was available for the project. Hence, the selection of tree species was mainly determined by this time pressure and the seeds available at the start of the project. More time should be invested in the selection of appropriate species and seeds.

The budget calculation for any tree planting project needs to take into account the necessary maintenance of the plantings for at least 1 year after planting. This includes irrigation, pest management, pruning and fertilizing.

The support and continuing involvement of all relevant stakeholders and a close interaction with the involved staff was however invaluable at all stages of the project (cf. Engel and Korf 2005). Supporting measures aiming at the rising of public awareness for tree conservation and planting measures should accompany similar projects.

4.5 Mass Flow Management

4.5.1 Starting Position, General Considerations

Water is one of the most important topics in Africa and at VVU, actually the whole program started because of water considerations: the request to visit VVU 2001 was about wastewater treatment there, but the first inspection made clear, that wastewater was not the problem, because there was scarcely any water available, and all water had to be brought by VVU´s own truck. But it was also clear, that a holistic overall concept was missing and also one for the mass flow management. In the course of the ecological development of VVU both have been designed (Fig. 4.31).

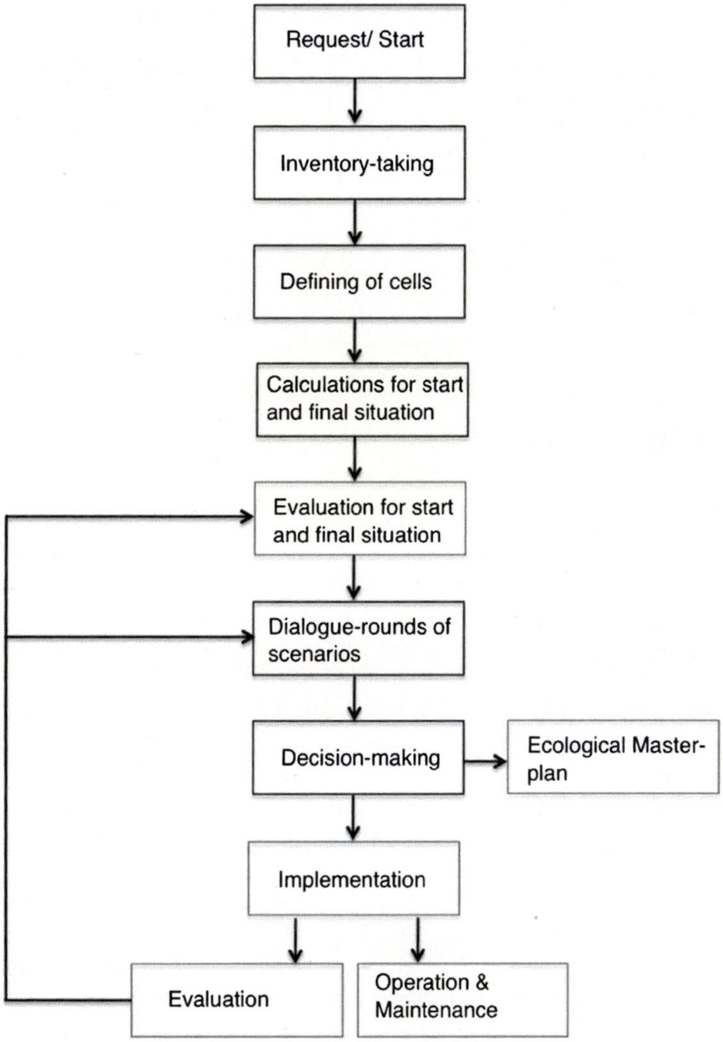

Fig. 4.31 Flow chart of mass flow management

4.5.2 Mass Flow Analysis

The first step in evaluating the system concerning cycles of matter must be a comprehensive analysis of the mass flows including the imports, exports and the internal cycles. That gives a first indication, where the system is unbalanced and vulnerable. In many cases of settlement ecosystems on their way to sustainability the flow of waters (rainwater, grey water, yellow water = urine, wastewater etc.) and organics (faecal matter, organic waste) is most important. Here the analysis must differentiate between the various producers and consumers, natural (men, animals) and technical ones (machines, buildings). In the case of VVU concerning

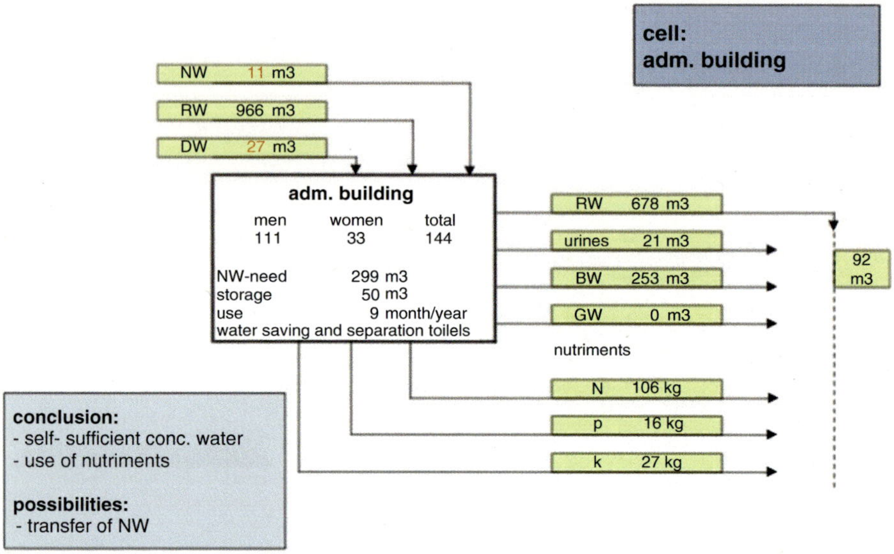

Fig. 4.32 Mass balance for faculty building of VVU

mass flow of waters various types of producers and consumers have been identified: producers of rainwater are the administration and faculty buildings and the cafeteria (their roof area can harvest more rainwater than is needed in this buildings), the dormitories are producers of greywater, the sanitary buildings of urine. Consumers of rainwater or service water are the dormitories and the agric and gardening area, the agric and gardening area are also consumers of greywater.

The following schemes show a faculty building and in contrast a dormitory and the mass balances of VVU with 1,000 and 5,000 people on campus respectively.

All cells together give the picture for the whole settlement-ecosystem, here for the VVU-campus. The input and output give an impression about the autarcy of the system and first hints for measures to improve the situation if necessary. For VVU it shows that water demand at the initial state with 1,000 people on campus ("inhabitants") can be secured under the condition that the supply is guaranteed. Projected into the future with wanted 5,000 people on campus the situation is quite different: the water supply could not be satisfied under the present conditions and kind of sanitary infrastructure. Important possible measures to improve the situation could be water-free installations like composting toilets and much more rainwater usage (Figs. 4.32–4.35).

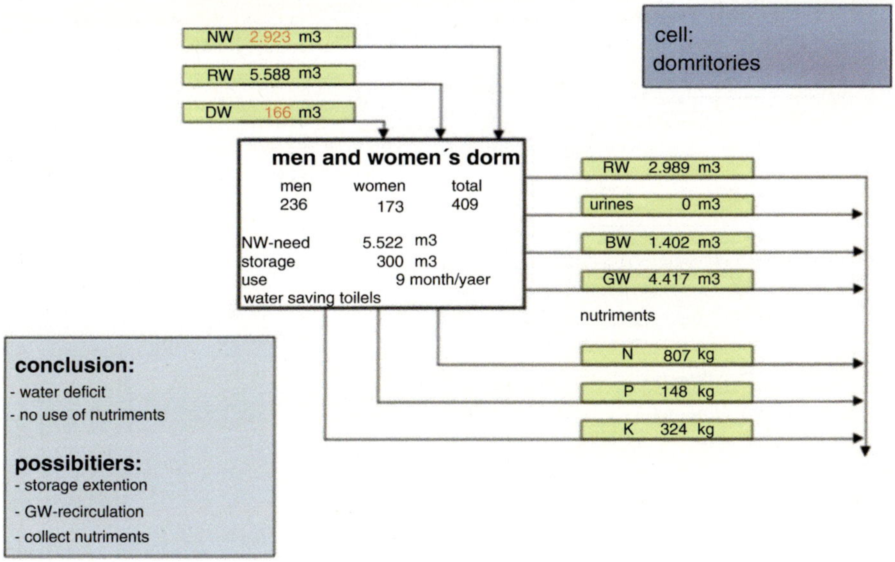

Fig. 4.33 Mass balance for dormitories of VVU

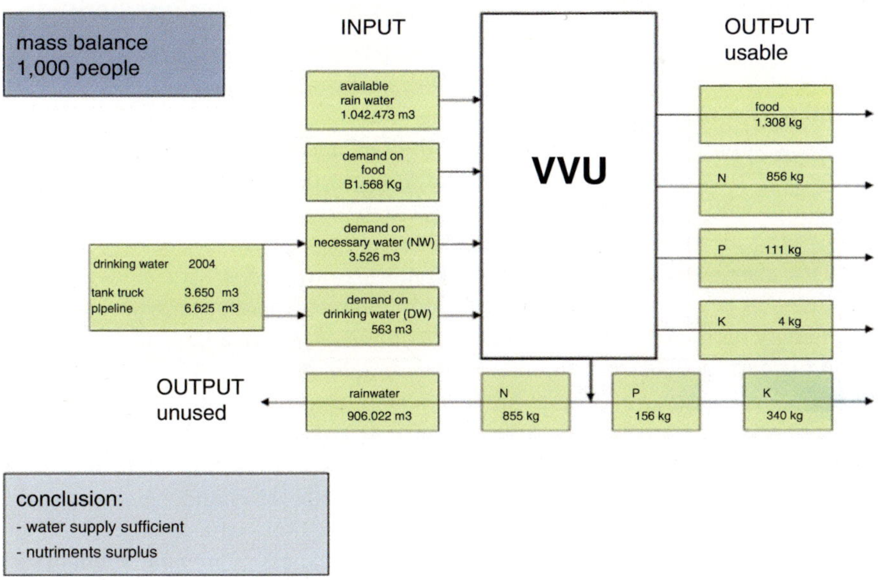

Fig. 4.34 Mass balance for VVU with 1,000 people on campus

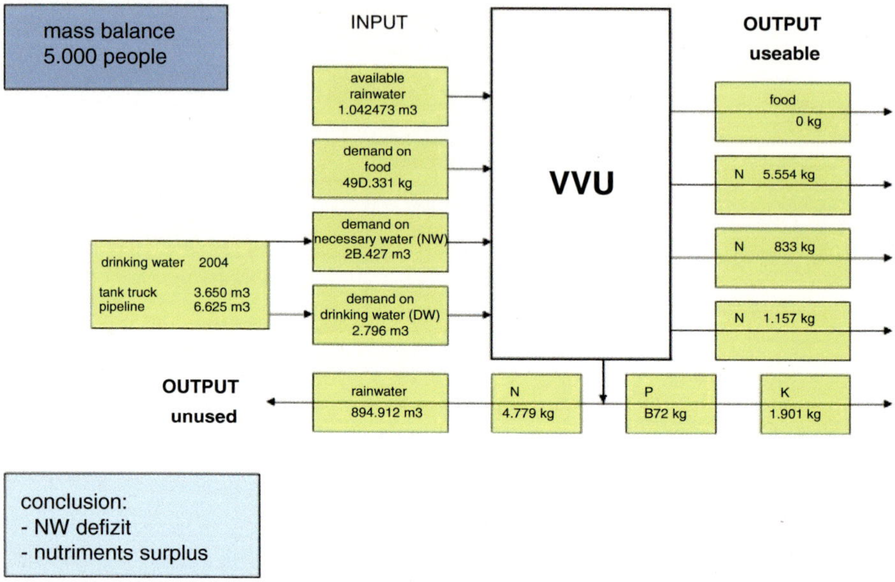

Fig. 4.35 Mass balance for VVU with 5,000 people on campus

4.5.3 *Integration of Mass Flow and Land Use Management*

From the start it was the intention to generate an instrument of decision making by computing the mass flows of the starting phase and those of the final stages, comparing various ways of sanitary installations and of organizing the buildings and the use of the space, including that for greens and agricultural use. The installations under consideration could have been more or less water saving, composting toilets being the most-water using units.

For the distribution and arrangement of buildings and green/agric areas two main scenarios had been chosen: the “centralized” masterplan solution with concentration of study activities in the center of the campus and with a big assembled agric area at the east and south-east corner of the campus (see Sect. 4.2.4) and a second “decentralised” scenario, where producers and consumers of waters would be closely connected.

For this scenarios the cell model and the ecosystem model with technical-cultural elements had been adapted, defining cells of similar characteristics concerning production and consumption of waters. From this point of view the dormitories are one kind of specific cells, producing excess of greywater, but in lack of rainwater for the needs of the users. On the contrary the faculty buildings because of their large roof areas are producing excess rainwater, far more than is needed for the toilets there. Another cell would be the green/agric areas, which generally always will have a deficit in water. Thus combining synergetically such complementing cells was the idea of the second scenario. However the positive effects of these combinations from the point of mass flows have to be balanced against the other criteria for the campus development: short ways for the users, quiet rooms for study, productivity of green/agric areas when assembled or dispersed etc. The dialogue about this matters enables clear understanding and decisions based upon that and the dialogue in itself improves the cooperation of the participants, the mutual understanding and the establishment of trust and even friendship.

4.5.4 General Guiding Principles

Ecological mass flow management must follow the general ecological principles. This means reducing unnecessary imports and exports, effective multiple use of matter in the (VVU-) System, closed cycles where possible, utilizing synergetic possibilities etc. This will result in relative autarky, which in human-based ecosystems like villages or cities never reaches a degree like in most natural ecosystems. But every step in this direction will help to gain balance between the needs of the human systems and those of the natural ones.

4.5.5 Main Findings

The lesson learned for future projects is to do a mass flow analysis even in the starting phase, which not only deals with the flows but also defines the specific cells in respect to their characteristics of consumption or production. In the following dialogue-rounds the possible combinations of cells in respect to the ensuing settlement and land use patterns and their respective advantages and disadvantages have to be discussed intensively. This allows that out of that there is a good chance to find the best solution for the needs of the local people and of the local ecosystem.

4.6 Management of Cycles/Water Management

4.6.1 Vision, Goals and Water Projects at VVU

The wish of VVU to develop into the first African ecological university is clearly stated in its ecological master plan. In the coming years the university is planning a steady increase of the number of students enrolled as well as of the university staff (5,000 students in 2015). In order to provide this population with a safe and continuous access to water supply and sanitation, the water services need to be developed.

In regards to water services, several projects have been carried out during the past years. These projects were focusing on different aspects of water supply and sanitation such as the reduction of the water demand, the development of rainwater harvesting, groundwater use and ecological sanitation. The main outcomes of these projects are described in this chapter. Prior to the outcomes of these projects, a brief presentation of the water supply and sanitation situation is presented for the reader to better understand the local context and challenges faced by various actors of the water sector in Ghana and at VVU.

4.6.2 A Brief Overview of Water Supply and Sanitation in Ghana

According to the WHO/UNICEF Joint Monitoring Programme (JMP) for Water Supply and Sanitation, 82% of the Ghanaian population had access to an improved drinking water source, whereas only 13% had access to improved sanitation in 2008. The lack of access to clean water and sanitation systems is a central public health concern contributing to 70% of the diseases in the country (AfDB/OECD 2007). At the national level, the total water demand is actually lower than the total amount of fresh water available, but there are strong regional disparities. In addition to insufficient water resources in some regions, the sector generally lacks the financial resources to operate, maintain and expand water infrastructure. AfDB and OECD further mentions that tariffs are not always adequate to ensure full cost-recovery. An evaluation of the Ghana Water SC concluded that almost one third of the water supply systems in the country were non-functional. In some areas the discontinuity of the water supply is occasionally caused by power supply outbreaks, which for example is the case at VVU.

4.6.3 Water Supply at VVU

4.6.3.1 Context and Challenges

In 2009 approximately 3,000 students were enrolled at VVU, among these, 500 were living in the campus dormitories. Additionally, some of the university staff as well as guests were also residing inside the campus (100–200). The total water demand was estimated to range between 70 and 90 m^3 per day in 2009.

At the time of writing, there are still many days during the year where water demand is not met. These water shortages are often due to the irregular supply of water from external sources, upon which VVU strongly depends. Another reason is the frequent power supply failures, which prevent the community from pumping water.

Today, there is still an urgent need to develop water services at VVU. If nothing is undertaken, the water shortages will become more and more frequent. This is a

Fig. 4.36 Water tower at VVU

serious challenge as the capacities of the available water resources and of the existing water infrastructures appear to be limited.

In the very first years of VVU's development, water was supplied to VVU's community by a truck equipped with a 20 m^3 tank. Along with the campus development, the water demand increased and this practice became non viable. Other solutions were needed. In the early 2000s, the newly constructed water supply system of Oyibi started to deliver significant amounts of fresh water to the university. The water was then distributed to all buildings via VVU's own network of reservoirs, pipes and pumps.

Although Oyibi Water Supply Scheme (OWSS) has been VVU's main supplier of water, its capacity alone is not large enough to continuously meet VVU's water demand. Consequently, VVU started to search for other alternatives. Rainwater harvesting and ecological sanitation were developed. Boreholes were drilled in the campus.

With the rapid development of the Greater Accra Region, OWSS has to supply an ever growing population while its water infrastructures are not being developed. Consequently the amount of water delivered by OWSS to VVU is very likely to decrease in the next few years, which could in turn seriously threaten VVU's development (Fig. 4.36).

4.6.3.2 Towards a Continuous and Adequate Water Supply

All projects have contributed to improve the situation of water supply at VVU. The projects have been focusing on both demand and supply sides of water management. The partners have tried to reduce the water demand by implementing water saving technologies and working closely with the community. Current and future challenges as well as possible solutions were discussed in detail. Subsequently, new water supply infrastructures were constructed resulting in the development of systems for rainwater harvesting, groundwater use and water treatment.

VVU's water supply infrastructure has consequently shifted from a centralised system toward a hybrid system, which combines centralised (groundwater supply) and decentralised subsystems (rainwater water supply).

4.6.3.3 Demand Side Measures

Evaluation of Water Demand

One of the first main steps undertaken was to estimate the water demand as it is essential information required to correctly design water supply infrastructures. Krämer (2007) estimated the water demand of students living in the campus dormitories. Depending on the installed devices in each building (e.g. flush or dry toilets), the water demand ranged from 60 to 100 L per capita, per day.

Metering of Water Consumption

Another key step was the metering of actual water consumption, which started in 2006. Weekly readings of the meters showed the distribution of water consumption at the campus. In this way, it could be identified that 70% of the water was consumed at the dormitories. It also showed that the water demand was not met and that the supply of water was not continuous.

Implementation of Water Saving Devices

Along with the collection and evaluation of water demand and consumption data, the buildings constructed during the projects were all equipped with water saving installations such as water saving shower heads, low flush or dry toilets. Before implementation, the question of acceptability and applicability of such water saving devices was investigated at VVU. Several water saving devices were tested and some were rejected because their maintenance would have been too extensive or because their energy requirements were too high for VVU. The acceptability of dry toilets was an issue in the beginning but discussion with students revealed that they would accept this type of toilet.

For long term sustainability, such water saving installations should be installed within the university, especially in the dormitories. It was estimated that through the installation of water saving equipment or dry toilets, the total water demand could be reduced by 15 and 20%, respectively. If both scenarios were combined, water use could be reduced by 35%.

Fig. 4.37 Low-flush toilets with urine separator (5 instead of 10 L)

Implementation of Wastewater Reuse

The introduction of wastewater reuse practices led to the reduction of fresh water demand. Black, yellow and grey water is still used to irrigate local crops (Fig. 4.37).

4.6.3.4 Supply-Side Measures

Rainwater Harvesting

Several rainwater harvesting systems have been implemented at VVU. Prior to their implementation, rainwater gauges were installed in order to collect rainfall data and estimate the potential of rainwater harvesting.

The measurements have shown that VVU campus receives about 700 mm of rainwater per year. Rainfall is not uniformly distributed throughout the year and rainy as well as dry seasons can clearly be distinguished (see Sect. 4.1) (Fig. 4.38).

Fig. 4.38 Rainwater harvesting at the guesthouses

The fluctuation in seasonal rainfall considerably limits the use of rainwater as a water source. For most buildings, the water supply cannot be based on the sole use of rainwater alone. Water harvesting surface and storage tanks are not large enough to store enough water to enable a continuous supply. Rainwater harvesting should not be neglected because of a lack of reliable alternatives, but rather be considered as an additional and complementary water source.

The most important elements of a rainwater harvesting system is the collection surface and, beyond all, the storage tank. Usually, the collection surface already exists or will be constructed anyway (e.g. roof). Storage tanks need to be built additionally and normally represent the largest investment cost (Thomas and Martinson 2007).

The size of the storage tank depends on the total amount of harvested rainwater, water demand as well as the temporal distribution of these two variables. It is therefore critical to collect reliable and sufficient information.

Roof rainwater harvesting systems were implemented at the staff and guesthouses, women's dormitory, new faculty building and Baobab Center. At the staff and guesthouses, i.e. the buildings with the highest collection surface (40 m^2) and storage volume per capita (5 m^3), it should theoretically be possible to harvest enough rainwater to meet the users' water demand (Box 4.5).

Two large-scale rainwater harvesting systems were constructed at the women's dormitory and at the Baobab Center in 2009. These systems differ from the others as not only the roofs but also large areas around the buildings are used to harvest the water. Each system has a storage capacity of 300 m^3, which is far above the capacity of the other tanks. Therefore, the systems are intended to supply not only the women's dormitory and the Baobab Center but also other buildings. Because rainwater flows on different surfaces, including ground surfaces before being collected in the storage tanks, it was recommended to filter and disinfect the water before consumption.

Fig. 4.39 Large underground rainwater tanks at the Baobab Center

Rainwater itself has a quality which is close to distilled water. However, the rainwater can become contaminated during the collection and storage stages. There are several pathways which can lead to a chemical and/or microbiological contamination of the water. Various materials, such as dust or animal droppings, can be found on the catchment surfaces and lead to the contamination of the water. Risk of contamination can be reduced by adopting measures such as the regular cleaning of the collection surface or the use of first-flush systems.

At VVU, analyses of the microbiological quality of harvested rainwater have shown that the water did not always comply with local drinking water requirements. *E. coli* and *Enterococcus* were found in some samples. Treatment will be necessary for drinking water purposes (Fig. 4.39).

Finally, rainwater harvesting systems must be regularly maintained (cleaning of gutters, reconnection of gutters or replacement of pumps, etc.). Therefore, the implementation of harvesting systems must also include sufficient human and financial capacities to manage these systems.

Box 4.5 Harvesting Rainwater at the Guesthouses

At the guesthouses, the entire roof surface is used to collect rainwater (340 m^2). The roof is made of cement tiles covered by a red pigment. Rainwater is collected in PVC gutters and pipes, filtered by a mosquito grid with a mesh size of 2 mm and then directed to a 42 m^3 concrete underground storage tank. Finally, it is then pumped into two 3 m^3 PVC overhead tanks. There are usually up to eight people living in the guesthouses. With a daily water demand of 80 L per capita, simulations have shown that the system is well designed to operate as an independent cell. The harvested rainwater is used for all domestic uses except for drinking purposes.

Groundwater Use

Since 2008, three boreholes have enabled VVU to extract large quantities of groundwater. In 2009, approximately 45% of the system input volume of the VVU central water supply system originated from these boreholes (i.e. about 500 m^3 per week). The boreholes thus represent the second largest source of water after OWSS (55%). One of the main reasons for drilling the boreholes was the fact that OWSS could not deliver enough water to VVU.

Compared to rainwater, groundwater has several advantages. First, its quality, especially for drinking, is normally higher than surface water as the soil layers provide better protection against pollution. It also contains minerals, some of which support healthy nutrition. The microbiological quality of groundwater is also usually much higher. Second, the storage capacity of aquifers is normally large although there is always a risk of over extraction. Over extraction should be prevented as it can have severe consequences in the long term (e.g. salt water intrusion in coastal areas, groundwater-related subsidence).

In 2009, measurements at VVU have shown that the groundwater level was approximately 40 m below the ground surface. However, at the time of writing, it is difficult to predict the evolution of the aquifer within future years. Current information appears to be unavailable. If the context remains unchanged, the number of private boreholes in the region may increase. This could lead to a reduction of the groundwater level in the region. In that case, VVU may need to deepen the existing boreholes or drill new ones for further water extraction.

The best way to anticipate a decrease of the groundwater level is to monitor the evolution of the aquifer. Therefore, measuring devices were installed in 2009 in each borehole. Their function is to monitor the evolution of the groundwater level and of the water quality (temperature and electrical conductivity).

Physicochemical analyses have indicated high iron and turbidity concentrations for some groundwater samples collected during the same year. These samples were not in compliance with the local requirements for drinking water. One of the main problems with iron and turbidity is that they can lead to the accumulation of deposits in the supply system, which can cause operational problems. Furthermore, iron can also impair colour, give a metallic taste to the water and stain clothes.

Groundwater Treatment

VVU community intends to use groundwater from their boreholes to produce drinking water. As the groundwater has not always been complying with the local requirements, it is necessary to treat it. Following discussions with the project partners, the implementation of small-scale treatment was decided in 2009.

The key components of the small-scale water treatment plant are two filtration units, which use anthracite as a filter medium. This material can be easily backwashed with a low quantity of energy. The advantages of these units combined with the long life span of the filtration material convinced the partners to opt for this

Fig. 4.40 Construction of a small-scale groundwater treatment plant

solution. The treatment plant was designed, planned and implemented in a comprehensive way, i.e. taking into account community, technical, environmental, financial and institutional aspects. Given the local context and especially considering local capacities, treatment processes based on the use of chemicals were excluded. The classical processes of aeration and filtration were hence selected to improve the water quality (Fig. 4.40).

Drinking Water Production

The projects have shown that one of the main reasons for insufficient maintenance and development of existing water infrastructures was the lack of local financial resources. As with other communities in Ghana, VVU encounters difficulties to recover the costs of continuous water supply and sanitation services.

Currently, the community has been purchasing bottled or sachet water for drinking purposes. Sachets are small plastic containments made of high density polyethylene, which are filled with drinking water. They are produced by private companies and can be purchased almost everywhere in the country. Compared to bottled water, the price of sachet water is significantly lower.

In past years, the VVU community has expressed the will to produce its own drinking water. The sales of sachet water to the community could enable VVU to raise additional funds for the maintenance and further development of water and ecological infrastructures. Therefore, a sachet water production was implemented during the climate change project.

The sachet water factory at VVU uses a similar production process as other sachet water factories in this region (Fig. 4.41). It consists of a storage tank, a distribution system (pipes), a decentralised water treatment system (filters and UV-C

Fig. 4.41 Sachet water production

disinfection unit) and a packaging system. The main feature of the treatment system at VVU is the use of a low-pressure membrane filter, which requires a low amount of chemicals and energy for water processing. Compared to the commonly used polypropylene cartridge filters, the unit offers better filtration performance and also produces less waste because the membrane can be cleaned and used several for years, whereas the polypropylene cartridges filters have to be replaced frequently.

Drinking water production should follow a strict quality management procedure to ensure safe and best water quality. It is planned that the plastic waste originating from the production and use at VVU will be collected for recycling.

4.6.4 Ecological Sanitation at VVU

Ecological sanitation is a holistic approach to sanitation and water management based on the systematic closure of local material flow-cycles (Werner et al. 2004). It recognises human excreta and household wastewater as resources which can be recovered, treated where necessary and safely reused in agriculture (Münch et al. 2009). Because of the nutrient content of domestic waste water, ecological sanitation can lead to the improvement of soil fertility, food security as well as to the reduction of water consumption and pollution of water resources.

From 2004 to 2007, many efforts have been made to develop ecological sanitation at VVU. Several systems have been implemented and analysed regarding many aspects such as water quality, impacts on crops and environment, cost effectiveness and acceptance. Some of the implemented systems are described below. Ecological sanitation products are now used for Valley View farming activities (e.g. mango cultivation).

For ecological sanitation it is necessary to distinguish between the different types of wastes/resources: urine, faeces, black water (a mixture of urine, faeces and flush water) and grey water (waste water generated from washing food, clothes dishware and bathing).

Each technology has its advantages and disadvantages, which must be considered during the selection process. Again, technical, environmental, communal, financial and legal aspects must be considered.

To collect, treat (if required) and safely reuse the waste water, different technologies are available. Tilley et al. (2008) define five main functional groups of technologies involved in sanitation:

- User interface technologies (e.g. dry toilets, flush toilets)
- Collection, storage/treatment technologies (e.g. twin peaks, storage tank, septic tank)
- Conveyance technologies (e.g. sewer, drains)
- Semi- (centralised) treatment technologies (e.g. aerated pond, constructed wetland)
- Use and/or disposal technologies (e.g. land application, irrigation)

4.6.4.1 Urine

Urine is collected for reuse at various buildings of VVU (cafeteria, administration and one faculty building). Urinal and/or diverting flush toilets are used as "user interface" technologies. The urine is then directed toward plastic tanks where it is stored. When the tanks are full, a tractor equipped with a motorized pump and a storage tank is used to pump out the urine and to transport it the crop fields (Fig. 4.42).

4.6.4.2 Grey Water

Grey water collection and treatment systems were implemented at the cafeteria and the women's dormitory. At the cafeteria, grey water is waste water originating from the kitchen. It is estimated that approximately 2 m^3 of waste water is produced every day (Palutec 2008). Grey water is collected in a large three-chamber storage tank, which is designed to remove grease and the largest particles. The grey water is used during the dry season to irrigate the nearby papaya plantation.

At the women's dormitory, grey water essentially originates from washing activities. The grey water is simply collected in a large underground storage tank and then pumped to the crop fields (mango, banana and avocado).

4.6.4.3 Black Water

At the guesthouses, water saving flush toilets are used. Urine, faeces and grey water are not separated. Black water is collected in an underground tank where the largest

Fig. 4.42 Urine collection tanks

particles are settling out. Waste water is then pumped to a filter made of laterite (nitrification and reduction of microbial loading). The black water is then used for gardening purposes.

Dry toilets at the men's dormitory as well as a biogas production facility located near the cafeteria are also part of the ecological sanitation concept, which was implemented at VVU.

4.6.5 Main Findings

The projects have shown the importance of background research, which is essential for the most efficient design of water infrastructures (e.g. research on water demand).

The monitoring of water infrastructures is also important for the evaluation of the infrastructure performances.

The experience with water saving installations was positive. Such technologies were accepted and significantly reduced the campus' water demand. In the context of an increasing population and unreliable water supply this is of utmost importance.

The development of ecological sanitation also contributes to the reduction of the campus' water demand and is a further benefit for agricultural activities on campus.

Rainwater harvesting is a good means to increase the amount of water available. However, it must be kept in mind that only in some cases (low water demand, high collection surface and storage volume) a water supply system can be solely based on rainwater harvesting.

For groundwater use, simple technologies were selected and successfully implemented.

A key aspect highlighted during the last project at VVU was that providing a continuous and sufficient water supply is not only a problem of limited water resources but also a problem of energy. Power shortages prevent the continuous supply of water. Using solar energy could be a viable solution. Pumps using solar energy will be used to pump water out of the new underground rainwater tank constructed at the new Baobab Center.

4.7 Ecological Design and Building

4.7.1 Ecological Holistic Concept

Ecological design and building is embedded on the one hand in the physical-scientific default of the environment and, on the other hand, in a process of the design (see Fig. 4.43). Thereby it is basically different from the used standard of design and building: Before the planning itself an integrative ecological concept over all must be developed with reference to the common standards – of course.

The architectural demands and wishes are fitted into the requirements of the environment. Synergetic effects can be used especially well if the ecological concept is developed for a big area.

By the many fields to look for and because of the multiplex architectural requirements it must be worked after certain method and systematically – in several flows from the start up to the goal and dwelling on branches and aspects (see Fig. 4.44).

4.7.2 Analysis of Situation and Conditions

At the beginning of the works there was a proposal for the future development which intended a park-like use of the area with 120 ha according to the model more

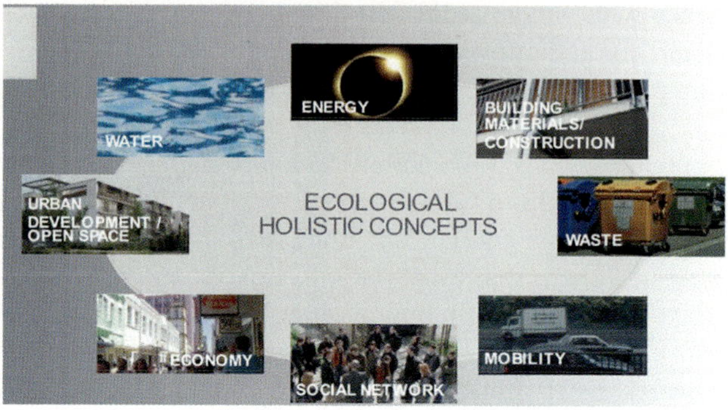

Fig. 4.43 Ecological holistic concept: many fields are part of the ecological design and building

prestigiously of American universities (see Sect. 4.2). The entrance flows on a big oval of approximately 700 m length and 250 m wide with the church at the beginning and the library into 700 m of distance at the end. Within the oval the different faculties and management buildings are placed, outside the hostels. Each building was reachable by car.

Between the buildings large green areas were provided. Large areas were wasted. After completing of the university long ways by a coach are needed which also drives on the oval. In this time long pipes were already installed. Ecological aspects were disregarded, the whole technical infra structure would not be financeable in the construction as well as in the use. The immense water consumption would lead soon to an emergency care, the climate control of the buildings would have closed big energy amounts winding, the ways and between the buildings are wide. Ecology was no special subject in this time.

With the complete new planning of the campus the first ecological university of Africa should be realised with a concept over all. Moreover a master plan Group was founded in the VVU with the most important people involved in the decision trial.

4.7.3 Common Goals

The common target or higher aim of the total concept is the "Stadtschaft" or "townscape", so a campus for apprenticeship and research in which at the same time the natural cycles are built up in particular for water and nutrients and which offers a high sustainable quality of life at the same time to the users. It should avoid possible risks like water shortage and energy shortage, high expenses for the infrastructure, excessive land use and decreased stay quality, e.g., by the motor traffic and heat.

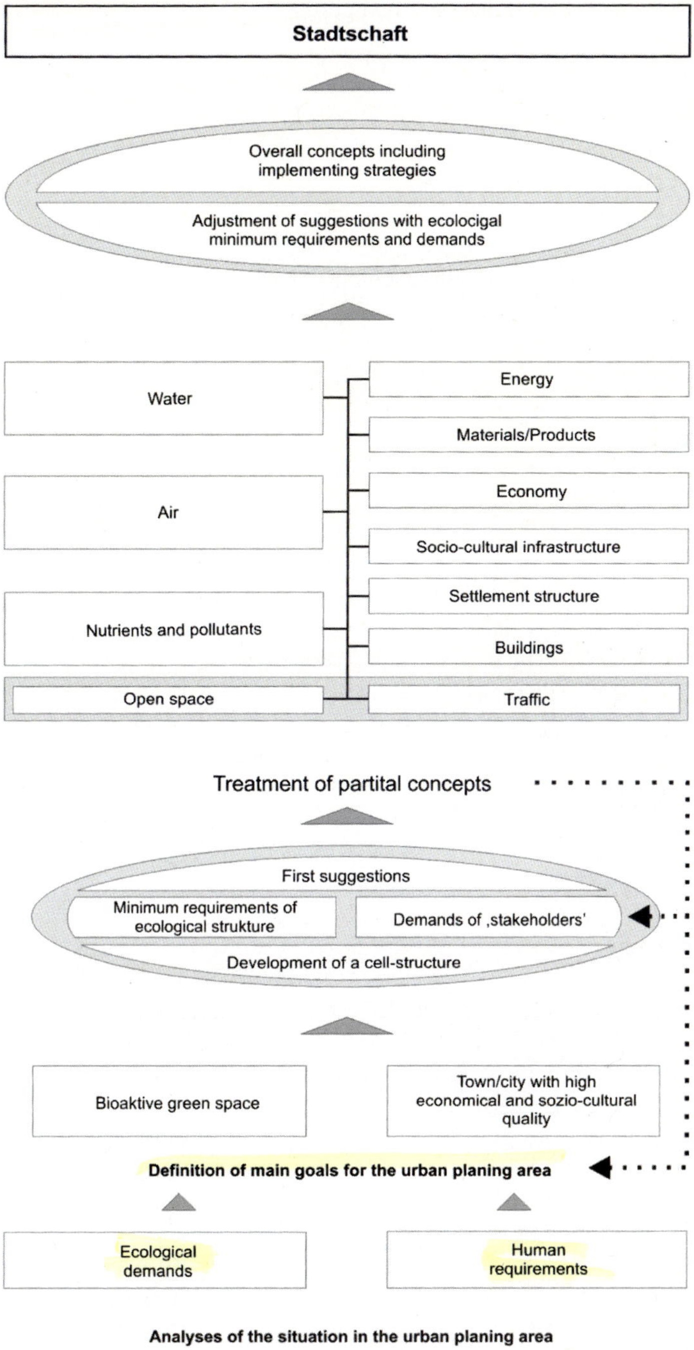

Fig. 4.44 Flow chart of ecological building

Moreover some sting points for the approach:

- Certified density with linked up use
- Campus of the short ways
- Car free campus
- Experience campus
- Multifunctional bioactive open space
- Cycling of water and nutrients
- Campus of the practical environmental studies
- Agricultural use links up with teaching
- Connection with the surrounding villages
- Natural building cooling
- Solar energy use, etc.

4.7.4 Partial Goals, Common Goals and Approach

An integrative ecological over all concept and the accompanying masterplan are very complicated things that can only gradually in partial fields be compiled – as mentioned in Chap. 3. The partial results flow in onto the master plan which must be compiled in many steps forward and back.

For the VVU the partial fields of Energy, water, materials, free space and agriculture were examined according to a comprehensive analysis and demand inquiry and the results were transferred in the building design. The result is represented in the master plan.

4.7.4.1 Energy

Goal: sustainable energy supply under use of the energy saving potentials.

Energy was not a main object of the researches. The energy production (here above all by the hydroelectric power plant of the Volta Dam): The installations on the campus were not treated thoroughly – here still a lot of potential exists.

The building climate and the production of electric electricity was the main focus of the work. Mechanical cooling with air-conditionings can be renounced to (except laboratories) by planning measures: Good ventilation in the buildings by across airing, buildings crosswise to the wind direction (narrow buildings), shadowing partly also by trees, vaporisation cooling by plants, development of special ventilation blocks, Green Classroom in the shadowed court of the buildings.

The BAOBAB Centre for Ecological Studies receives a PV plant and becomes with it a building which produce more energy than use (negative CO_2-emission). A biogas plant in connection with the canteen serves for demonstration.

4.7.4.2 Water/Nutrients

Goal: An extreme water saving and nutrient use is reached by close-down water cycles.

The water supply is based on three measures: Rainwater use, own bore holes, connection with the public water supply.

Solutions: Extreme water saving measures by the installation, rainwater use with small and seasonal tanks, grey water separation, separating toilets and compost toilets use, grey water and urine use in the agriculture, compositing of feaces.

4.7.4.3 Materials

Goal: Cycling of materials and pollutant avoidance.

This branch was not worked extended, because it was not an object of the research. Green rubbish is composted and used agriculturally, plastics are collected and when recycling property is used, waste container on the whole campus.

4.7.4.4 Green Space and Agriculture

Goal: Green space after the demand of the "Stadtschaft"

Use of the green space to increase of the bioactivity of the campus: Wild plant zones, cultivations of plant communities and useful plants; additional use: Cooling of the open space and the buildings, water retention, space to cultivate useful plants, space for experience and learning (botanical garden, green foot path around the campus, special plant communities, rare Ghananian plants).

4.7.4.5 Mobility

Goal: Comfortable mobility with as low need of space and low environmental impact, high quality in use.

Qualified site density, campus of short ways, car free campus, only service vehicles, parking lot in the entrance area, campus bicycles, high stay quality by plants

4.7.5 *Cell Model: Shown by the Examples of Waters and Nutrients*

The complexly task of the ecology concept need a method to work on. In a first step spatial and functional cells are to suc for developing solutions – oriented in the campus (spatially) and in the function (partial fields). Inn the next step the cells are be to t linked up, linked up in the best way but not extremely optimised, for single parameters; since then the whole system is adaptable not enough. Then the solitary cells always contain a solution for itself and in addition with adaptation in the total concept (example see Figs. 4.45, and 4.46).

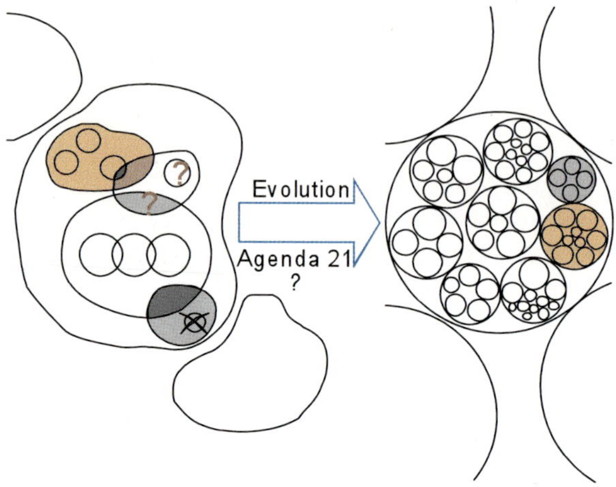

Fig. 4.45 Principle of cell model

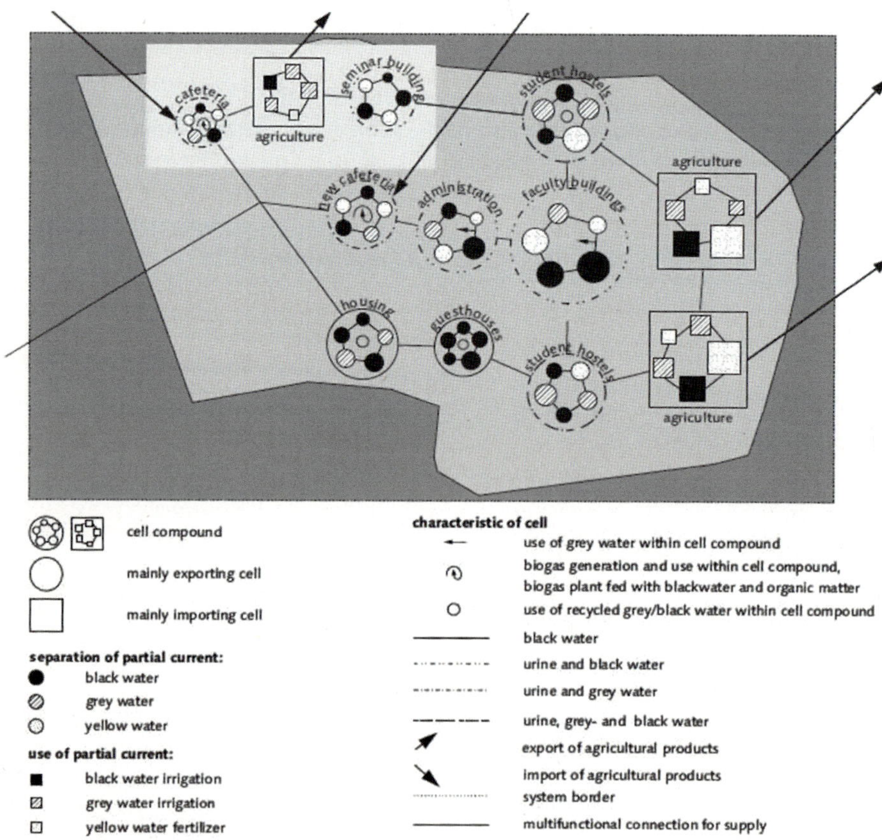

Fig. 4.46 Cell model for water cycling

4.7.6 *The Master Plan and His Architectural Realisation*

The master plan shows one of many solutions. To all the long trial of the search for solutions and the adaptation of partial solutions is common – often a compromise. In the concept proposal for the VVU leads in the master plan that gives the spatial use and the architectural development.

The master plan of the VVU shows the existing street oval and the access road. Along the public street and the entrance activities are settled like trade, sport and hospital which must be reached by cars and have a lot of visitors or where bigger events take place (sport). In the interface to the campus park all private cars. From there foot paths lead to the campus. The church with 2,500 places is placed as a dominating and symbolic way as starting point of the axis along the inner campus until farm area campus. The oval takes the service transport (Figs. 4.47–4.52).

Close to car parking the administration buildings are placed around the inner campus with different public activities. Canteen, lecture theatre and library centre for ecological studies etc. Afterwards the faculty buildings line up themselves along a generous promenade in groups (cells) with labs and auditoriums. The view goes from the church crossing the campus main place along the promenade between the faculty buildings to the open space with the agriculture.

Beyond the oval the residential buildings are placed in short footpath binding in the campus as a sort of village (residential cell). A little bit apart lie the guest houses with own access road. For every group of buildings is looked as a cell. A specific solution was developed, e.g., a decentralised water cycle system. In addition, they are fit in the over all concept, grey water from the Studentenhostel is led in the agriculture. The whole campus is circled with a teaching and experience path which begins and ends on the campus place.

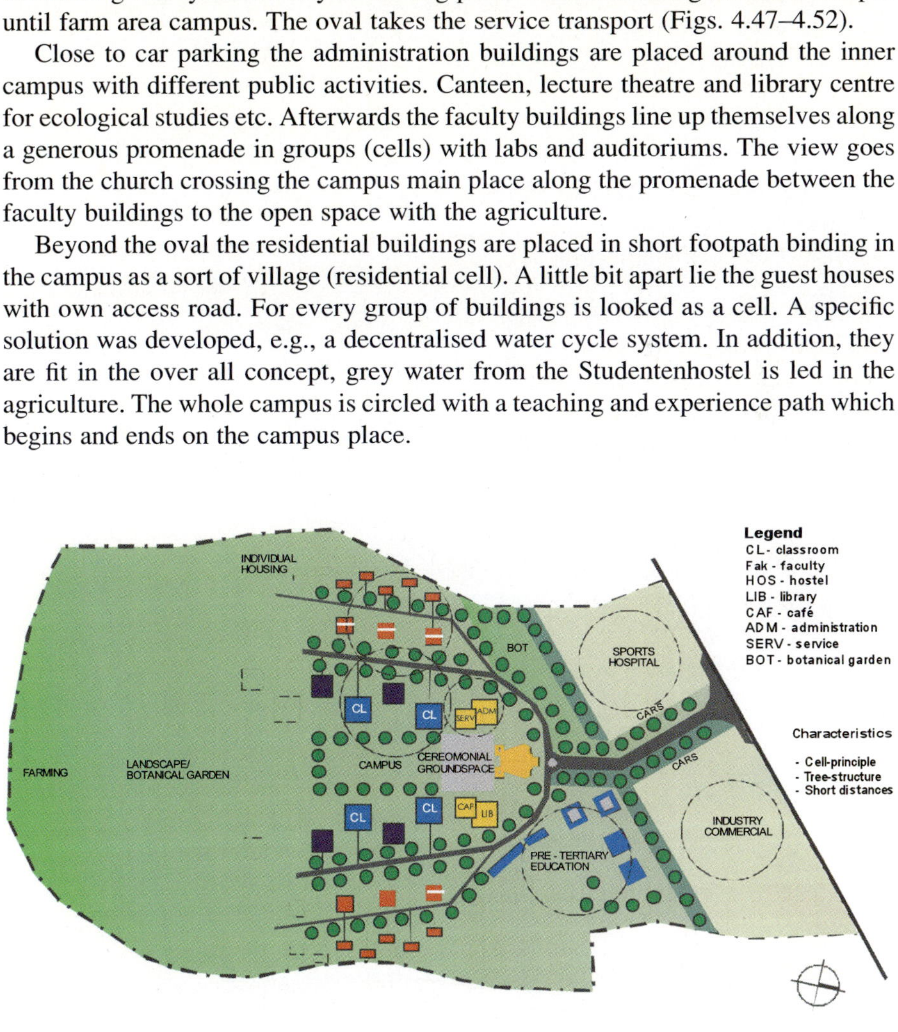

Fig. 4.47 Masterplan – first concept

Fig. 4.48 VVU masterplan of the integrated total ecological concept

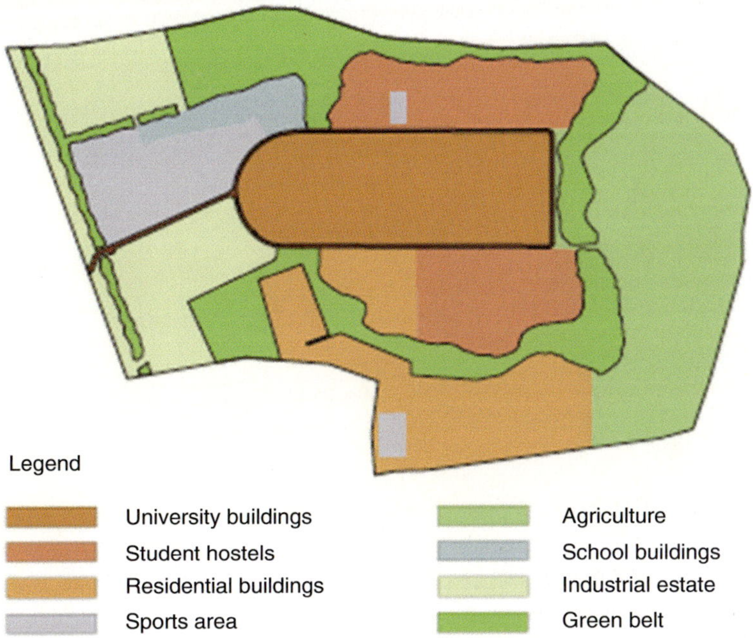

Fig. 4.49 VVU masterplan – partial concept: zoning of utilization

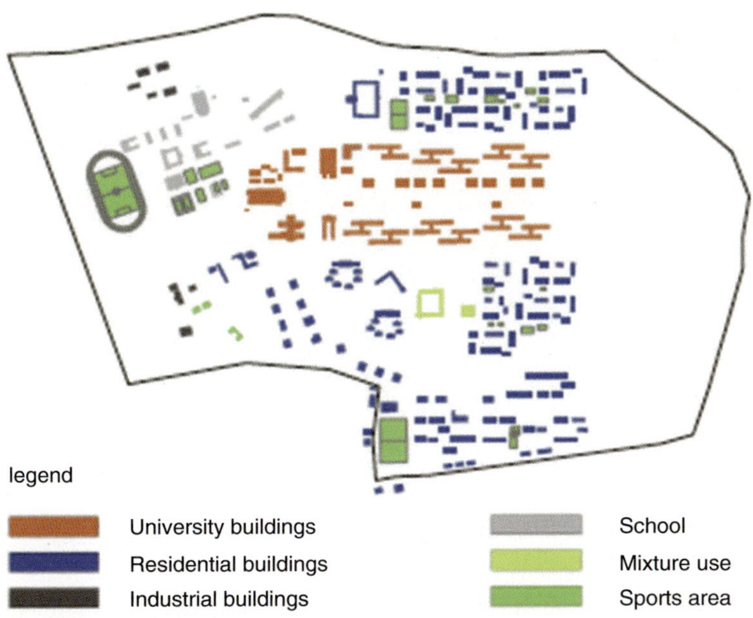

Fig. 4.50 VVU masterplan – partial concept: functions

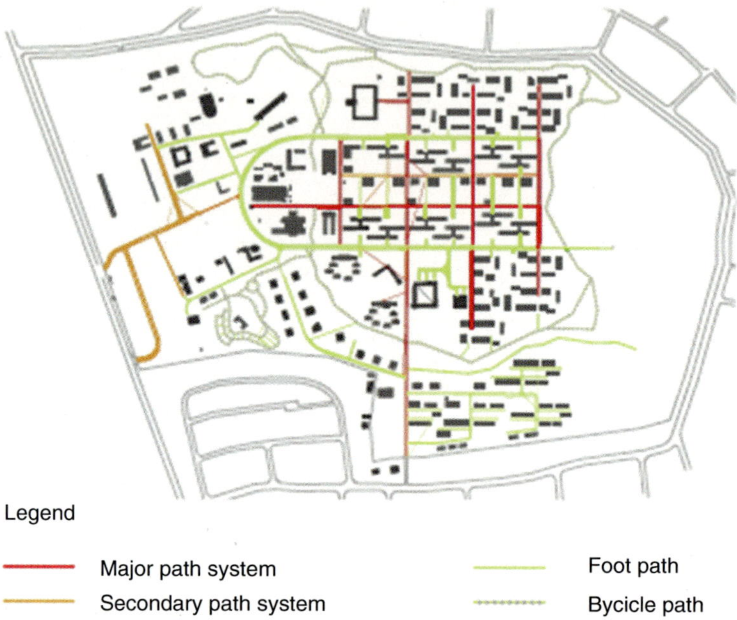

Fig. 4.51 VVU masterplan – partial concept: path network system

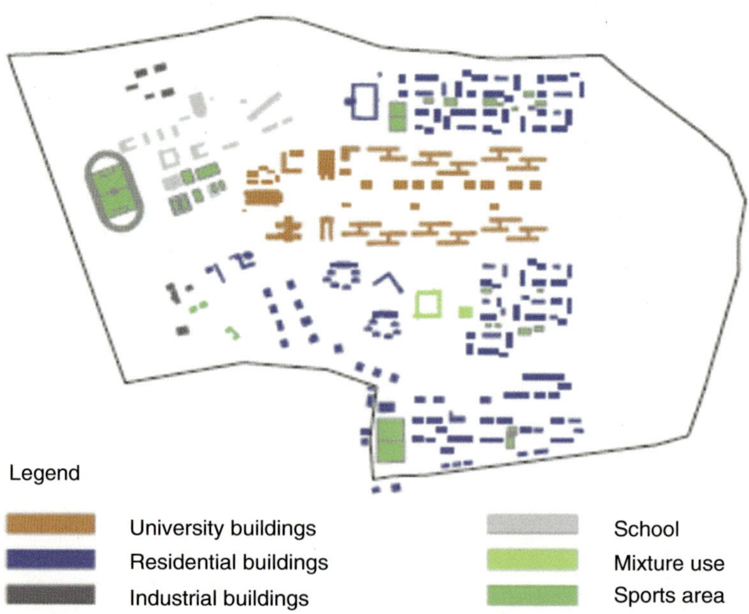

Fig. 4.52 VVU masterplan – partial concept: road system

Fig. 4.53 Sand filter for decentralized wastewater recycling

Fig. 4.54 Faculty building with ventilation system

All buildings are arranged crosswise to the main wind direction. Only one room is placed in the width of building for natural ventilation. Other buildings have special ventilation systems like ventilation blocks and chimneys (Figs. 4.53–4.55).

Up to now two large faculty buildings, the cafeteria, the BAOBAB Centre for Ecological Studies on the campus place, guest houses, a small hospital as well as several buildings for small industry are finished. Besides, there are installations to the water cycle, special sanitary blocks, the biogas plant, compost toilets, a lot of tanks for rain water and grey water and a sand filter to treat waste water. The agriculture and the whole plantation of the area is reported in other chapters.

The ecological over all concept an especially inexpensive and sustainable solution for the development of the university is reached. By the qualified density,

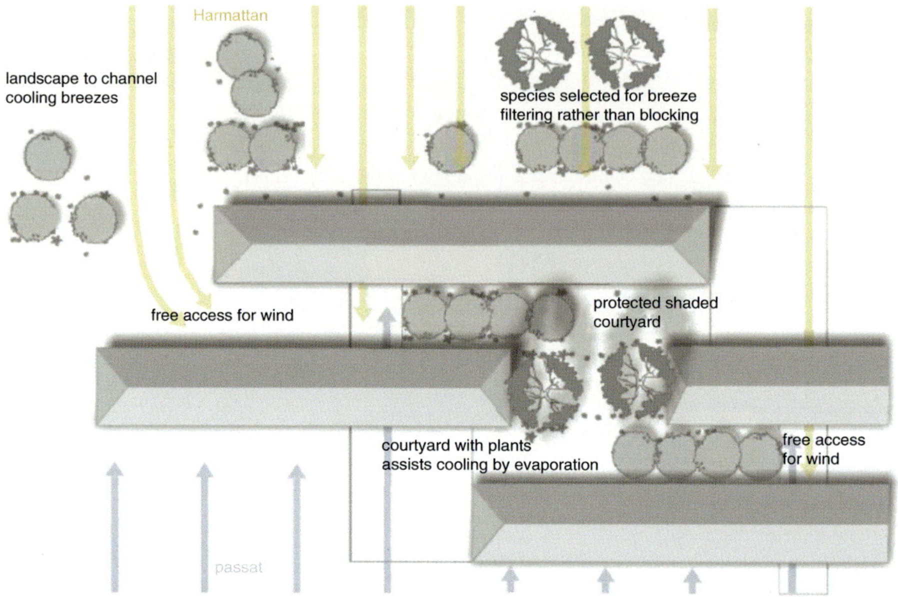

Fig. 4.55 Faculty building with ventilation system

the carfree campus, the measures for saving energy and water and the decentralised waste water treatment and reuse etc. the whole infrastructure and its management less expensive and more economically. Less pipes, less valves, less slots and less space are needed – namely with sustainable effect – compared with customary solutions. Rising resource expenses hit the VVU by far more slightly than conventionally planned equipments. At the same time the quality to stay and walk in the open space is especially high.

4.7.7 Main Findings

VVU is an example for the solution of an as an ecological campus.

It shows that such ecosystems can be planned and built under the sociocultural, climatic and economic conditions in a simple and functional way. Planning and building after an ecological over-all-concept is possible, provided that the stakeholders regard the recommendations of Chap. 5. Another precondition is their knowledge and experience of developing ecological over-all concepts.

Chapter 5
Lessons Learned/Relevance

5.1 Factors Influencing Ecological Projects

5.1.1 Cons: Factors to be Considered

The project-side in Ghana bears certain challenges which very often are beyond controllability by the project partners. Tropical climate conditions in combination with lack of national infrastructure may produce a shortage of water and/or power. On the other hand cloudbursts and excessive solar heat can stop or slow down construction works.

Infrastructural deficits of the country often delay deliveries or spontaneous purchases. In Ghana qualified craftsmen nearly are not available. This is true even more for the specific ecological program at VVU. Combined with the complex approach, which necessarily spans the whole chain of realization, a wide range of parameters have to be considered and put into operation, before the long-term functioning of the program can be reached. For the ecological approach the standards are higher than normally, for example concerning the buildings and installations. To give justice to the personnel, its framework has to be taken into account including its possibilities to keep up with its duties. One example may illustrate that: a contractor at the construction site relies on five skilled masons working steadily for him, five stand-by masons, ten labour men being hired on a daily basis. He has to provide transport for the workers to and from site, has to ensure weekly payment and has to provide tools. Eventually this can be one of the factors, which can produce hindrances and delays.

Sometimes cultural and social differences between partners provoked misunderstandings and offending actions. Additionally the mutual exchange of knowledge is limited, because cultural backgrounds are ruling. Views regarding sustainability and priorities in life are sometimes very contrasting between European and African partners.

For programs, like Research & Development ones, which deliver only small financial support for investments, education, training and operation and maintenance, some hints are important:

G. Geller and D. Glücklich (eds.), *Sustainable Rural and Urban Ecosystems: Design, Implementation and Operation*, DOI 10.1007/978-3-642-28261-4_5,

Without the unsolicited planning work of the designers and planners, which necessarily had to go far beyond the financed amount meant for designs, a qualified and ecological implementation in the intended scale hardly would have been possible, especially not in the necessary extent and time-frame. A second indispensible support for the long-term success of the project has been additional experts, like for the first R&D-program, where they have not been part of the original funding. Because they are decisively necessary, it is worthwhile to undergo the additional struggle to persuade the respective institutions for getting their financing.

Concerning the specific project-site and institution, the experiences can be summarized as follows:

To operate along a detailed and long-term implementation plan at VVU was not easy and caused losses of money and quality, because buildings and installations are established step by step depending on the incoming donations and funds. The many modifications of the approved designs caused a lot of additional work (unpaid) for the design team. Part was due to the always limited resources (man-power, qualification, rooms and equipment) and competences of the physical plant department.

The university owns a rather complex administrative network with distinct responsibilities and controlled scopes of action for different positions in the network. Complex bureaucracy is an obstacle for running a tight time schedule and in this project it needed to be slimmed in order to meet the essential deadlines. Cash-flow means progress in works and if there is an obstacle for cash-flow it has to be addressed immediately.

Negative influences on the time schedule were mainly cash-flow deficits, transparency in ordering details, organization of workforce according to demand on site, availability of workers (numbers), availability of workers (skills), availability of building material and resources (fuel/gas/water/power), bad quality of works resulting in demolition and re-building.

5.1.2 Pros: Supportive Factors

The ecological development at VVU was able to start under some very favourable conditions:

- Ghana was and is a peaceful, stable and not too poor a country
- The president of VVU was very positive-minded to the ecological cycle management as part of a holistic ecological development concept.
- The decision process at VVU was fast.
- VVU was in the starting phase of the campus-development, open to a lot of possibilities for ecological development, including town planning, ecological buildings, study of ecological engineering etc.
- VVU as reputed university could serve as positive example and good multiplier of concepts and experiences.
- The results and practical experiences open the opportunity, to be used by outside institutions, firms, schools and universities, which are involved in similar

projects or are discussing this kind of subjects in their respective fields of study, training or research.

It was favorable to start the ecological development with a prephase, which made it possible to intensively discuss and establish an ecological holistic concept and have experiences with a first small bunch of measures, before implementing the more extensive measures and installations of later steps.

The specific combination of the partners (universities, companies) covered an integrated approach (architecture, nature conservation, water management, agriculture, technologies, training, studies, quality management, public relation...) and proofed very helpful to bring the ecological development into realization and giving hope of their long-term functioning. The basic rules were set by the ecological masterplan providing the overall concepts for traffic, infrastructure, buildings, and land-use.

In order to achieve the objective of the project an interdisciplinary team of specialists was formed and all stakeholders in connection to the university were integrated into the development process. During the period of dreaming, planning, constructing as well as proudly celebrating the Grand Opening every participant was considered. The group of participants offered a broad variety of experiences, views and skills.

By listening and considering those during the process it was possible to analyse hindrances and problems properly and find adequate solutions. Interacting within different social and cultural backgrounds led to an exchange of knowledge into all directions. The resulting development has evolved from numerous discussions among the involved people and teams. With regard to the discussed matters re-considerations of economic, ecological and socio-cultural factors took place. Compromises were found where necessary in order to find the best solutions.

All stakeholders were motivated and dedicated to achieve the same goal, which was a fundamental prerequisite for the success. Without exception all parties were willing to communicate and a balance of conceding and insisting on planning measures can be recorded.

Furthermore it was especially fruitful, that the designs of the ecological masterplan, of the new buildings and of the ecological installations have been delivered nearly free of charge to VVU, thus making progress easy, fast and professional.

5.2 Holistic Planning in Practice

5.2.1 Integrated Land Utilisation and Cycle Management

The ecological cycles are implemented closely linked with the development of the campus-utilisation, including the settlement with its buildings, the greens and the agricultural area. This is important and needs special consideration due to its long-term effects, which are ranging over several decades.

This limits short-term experiments on the campus, as far as buildings and installations are affected: mistakes here can hardly be amended at all and often only by investing a lot of time and money. The integrated holistic development of settlements, the proper management of fluxes of matter and information however only can be verified in the field, not in the laboratory. Thus truly there is no alternative to the chosen approach at VVU.

The good thing is that now for the ecological development field-tested tools are available, like the "Stadtschaft"-approach and the cell model, developed for complex tasks in the field of town planning and for the first time tested on such a scale on the Valley View-campus.

Human ecosystems which involve cycle management are very often characterized by relatively high logistic standards and high requirements for technical equipment. Both are subject of specific planning and more than everything well-organized maintenance.

From our experience in Ghana the concepts with high consequential costs and efforts due to maintenance have very low chances to sustain. Technical equipment such as pipes, valves, filters and everything involving electricity like pumps, monitoring systems, etc. create major challenges for maintenance crews and financing of those.

To avoid failures the design of the project should be least difficult to maintain and to repair. Moreover there are constructions which respond in a certain obvious way in the case of damage (e.g. visible overflow of septic tank). Rather than disappearing invisibly in a soak-away in the underground we prefer the unpleasing but alarming ways which call attention to a fault.

5.2.2 *Integrated Information and Quality Management*

Integrated information and quality management was part of the holistic approach at VVU from the start. In the course of the ecological projects at VVU the scope of this part had to broaden and deepen considerably. The importance of the socio-cultural background for the realization of the ecological development showed up as essential in many respects, for communication, understanding, decision-processes, acceptance etc. The necessity of real understanding of needs and the socio-cultural background made it necessary to test and develop adapted methods and tools, some of which are presented in this manual. Fortunately many of the approaches and tools could be tested positively and may be of help in future projects.

Generally the importance of information in ecosystems and in ecological engineering (as communication, acceptance, organizational structures, socio-cultural background, communication, education etc.) seems to be grossly underestimated and needs to be an outstanding topic of ecological projects.

5.2.3 *Integrated Ecological Building*

The building design in the VVU was carried out up to now in usual manner: According to the demand a first design is done. After it's acceptance it is formed out to make it suitably'. Mostly it is modelled after historical creation or with such ornaments. Modern buildings like offices are designed after solutions of the of the north hemisphere with air conditioning that is a status symbol. Former experiences of traditional African buildings and examples from the colonial age are used a little. Simply and adapted solutions are considered as uncomfortable.

The project has shown that an ecological design is possible and it can be successful. The use of the buildings in the VVU Project and her surrounding is accept as positively, the building costs are very low. However future activities in this way are only possible if the acceptance is given. New projects should base on this, this quality is to claim before starting of the design of a project. It should be part of the tender documents. We need specialists who are trained for this. In this time it still happens in cooperation with a specified office mostly outside of the country. Constructions by ecological over-all-concepts as a basis of all decisions is hardly represented worldwide and so in Ghana not.

A corresponding international course of studies is necessary.

The first success could be reached for the tender for an administration building financed by the IMF.

5.3 General Conclusions

Two prerequisites are the foundation for a sustainable development:

- Consideration of all fields of sustainability (economic, ecological and social)
- Participation of all people involved

During the programs we learned that flexibility helps to obtain the goal. Not only during the design stage also during construction phases the design should be open for "forgotten" aspects. Also changing situations like national shortages of resources demand flexibility to keep up with quality and time schedule. Despite of all flexibility it must be clear that there is a guiding principle which has to influence the direction of all decisions.

The reasons for discrepancies have to be understood thoroughly before agreeing on the compromises. For a long term success of the facilities there is no way around the cultural and social wishes and a profound understanding of the behavior of the users. In general there is not an endless flexibility of adults to refrain from their ways of living and working. Compromising has to be balanced out between the parties for justice and acceptance.

Quicker administrative solutions are to be found and were found by outsourcing of money and procurement of material. In this way the financial and quality control could be secured. Reaction times for responsible persons within the hierarchy of university administration had to be shortened and the project had to be prioritized among other university developments. Key positions with dedicated and educated people are of utmost importance for a sustainable development in Ghana. These persons have to be installed with enough power to be able to take the necessary decisions quickly enough.

For larger projects it is essential to direct and organize the line of communication. Key positions and reliable people have to be identified. At the same time hierarchies are important and have to be acknowledged. If there is a way to hold people responsible for their duties it could be a way to keep quality and time frames.

Concerning operation and maintenance it is helpful to provide a certain budget as well as seminars for the operation and maintenance personnel.

Projects have to be designed for local circumstances. Climate factors have to be taken into account as well as availability of materials and feasibility of techniques. If possible low maintenance and long lasting constructions are the best choice.

The unique feature of the ecological development projects at VVU is their holistic and integrated ecosystem approach and the implementation of the measures considering all of the steps of realization, from first discussions over design and implementation to operation and maintenance. The flexible reproducibility of this example and its measures may be helpful to implement more sustainable solutions at VVU and elsewhere.

The ecological development of VVU is on a good way for the time being, despite some difficult circumstances. The testing of the integration of the ecological cycles into the holistic concept (masterplan etc.) has delivered valuable insights for future projects.

The various accompanying measures of a holistic information and quality management, like retreats, workshops, training, study, gaining acceptance inside and outside, additional personnel (CIM...) emerged as essential and turned out to be very supportive. Time is a key factor. This means for building up a reliable working relationship at least 5 years of cooperation are necessary, and many hours of communication, of meetings and working together, summing up to at least many month for each of the main stakeholders.

Normally specialists and experts involved in the project leave the site after a certain amount of time and a team of responsible trained locals have to take over responsibility. The following years are crucial for the surviving of the project and there should be seminars and educational programmes held for users and responsible teams during that time.

After nearly a decade of testing at VVU a lot of information is available about the necessary framework and other prerequisites for ensuring the establishment of a holistic ecologic concept even in the long run. The positive experiences with the holistic approach, the tested measures and the achieved results may be a valuable contribution to a sustainable development of settlements elsewhere on the globe.

Chapter 6
Authors, Partners and Institutions

6.1 Authors

- Geller, Gunther, Dipl.-Ing.: Ecological Engineering Society
- Glücklich, Detlef, Prof. Dr.-Ing.: Office for Sustainable Design
- Otte, Annette, Prof. Dr. Dr.: Division of Landscape Ecology and Landscape Planning, Justus-Liebig-University, Giessen
- Perfler, Reinhard, Dr.: Institute of Sanitary Engineering and Water Pollution Control, University of Natural Resources and Applied Life Sciences, Vienna
- Richard, Laurent, Dipl.-Ing., MSc: Institute of Sanitary Engineering and Water Pollution Control, University of Natural Resources and Applied Life Sciences, Vienna
- Simmering, Dietmar, Dr.: Division of Landscape Ecology and Landscape Planning, Justus-Liebig-University, Giessen

6.2 BMU-Project Climate Friendly VVU

6.2.1 Institutions

Federal Ministry for the Environment, Nature Conservation and Nuclear Safety (BMU), Bonn, Berlin
Valley View University (VVU), Accra, Ghana

6.2.2 Partners

Addai, Solomon BSc.: Agricultural Director, Valley View University, Accra
Bergmann Christiane, Dipl.-Ing.; Broberg Consult, Accra

G. Geller and D. Glücklich (eds.), *Sustainable Rural and Urban Ecosystems: Design, Implementation and Operation*, DOI 10.1007/978-3-642-28261-4_6,

Detlef Glücklich (DG), Prof. Dr.-Ing., Office for Sustainable Design, Buchholz
Holdefleiss, Antje, Dipl.-Ing., MSc, Berlin
Kwandahor, Emmanuel A. MSc: Physical Plant Director, Valley View University, Accra
Laryea, Seth, A., Dr.: President, Valley View University, Accra
Osei, Joseph K., Prof. Dr.: Provost, Valley View University, Techiman Campus

6.3 BMBF-Project Ecological Cycles at VVU-Campus

6.3.1 Institutions

German Federal Ministry of Education and Research (BMBF), Bonn/Berlin
Valley View University (VVU), Accra, Ghana

6.3.2 Partners

Addai, Solomon BSc.: Agricultural Director, Valley View University, Accra
Addy, Louis, Dipl.-Ing, CIM
Berger Wolfgang, Dipl.-Ing., Berger-BiotechnikGmbH, Hamburg
Fries, Nicola, Dipl.-Ing., Bauhaus-University Weimar
Geller, Gunther, Dipl.-Ing., President, Ecological Engineering Society IÖV
Germer, Jörn, Dr., Universität Hohenheim, Stuttgart
Glücklich, Detlef, Prof. Dr.-Ing., Bauhaus University Weimar
Höner, Gunhild, Dipl.-Ing., Palutec GmbH, Augsburg
Kwandahor, Emmanuel A. MSc: Physical Plant Director, Valley View University, Accra
Laryea, Seth, Dr., Prsident Valley View University Accra
Mitterer, Anton, CIM
Sauerborn, Joachim, Prof. Dr., Universität Hohenheim, Stuttgart

Chapter 7
Tools for Holistic Management of Human Ecosystems

7.1 The Tools and How to Use Them

There are a lot of tools available to make it easier ensuring a real holistic and at the same time sustainable, long-term functioning development of human ecosystems, especially for urban and rural settlement ecosystems. This toolbox encompasses diverse instruments, which serve the diverse needs and areas of operation.

The first ones are the process descriptions, which give an overview and illustrate the most important steps in the process of realization, the involved stakeholders for the different steps, the input for these steps and the results of each step. For the input also other tools to be chosen are listed, checklists or forms for example. To give a vivid graphic overview flow charts can be used (in our case preceding the process descriptions).

Checklists are tools for an easy way of controlling if all important topics are looked for, made for simply ticking off the listed topics, without too much pondering and explanations. This makes it a perfect tool for many cases, in which there is not much time available. Which is the case in many projects during the phase of realization and implementation of measures.

Forms are tools which give a structured way of answering important topics, delivering more information and insights, but also demanding more time to be filled. Questionaires are such a kind of forms.

Operation guidelines give elaborated information about procedures and especially for operation and maintenance. They have to be designed specifically for the intended purpose. It is important however to think of the users and prepare the manuals in such a way that every intended user easily can understand the information. It also should concentrate on the most relevant things: practitioners usually don´t take or have too much time for studying.

For each of the sub-topics (building, vegetation management, water management...) examples of the various tools are given. Those shall serve as examples, which should be adapted and modified to serve the respective needs best as possible. In Sect. 7.4 various QM-tools for the area of ecological building are

G. Geller and D. Glücklich (eds.), *Sustainable Rural and Urban Ecosystems: Design, Implementation and Operation*, DOI 10.1007/978-3-642-28261-4_7,

given, in Sect. 7.5 those for vegetation management etc. All tools are available in the original word-format, which makes it easy for use and modifying to the specific purpose (see www.ioev.de).

For the tools the following abbreviations are used: Process descriptions (PD), flow charts (FC), checklists (CL), forms (FO), Operation guidelines (OG).

7.2 Tools for Integrated Ecosystem-Management

7.2.1 Checklists

Checklist Sustainability Concerning People

Project:____

Responsible:____

		negative	*positive*	*neutral*	*Remarks*
People					
	Diverse outputs (productivity)				
	Food security				
	Yield (efficiency)				
	Risk				
	Income or income distribution				
	Capital requirements				
	Economic return, profit margin				
	Labour requirements				
	Maintenance / learning requirements				
	Self-reliance (uses local materials?)				
	Control over output and process				
	Living conditions (e.g., shelter)				
	Human health (e.g., sanitation, toxicity)				
	Energy supply (e.g., wood, fuel)				
Social self-determination					
	Family structure				
	Gender roles				
	Population growth				
	Education				
	Local institutions				

		negative	*positive*	*neutral*	*Remarks*
	Local culture				
	The rights of local communities				
	Community health				
	Local economy / capital flow				
	Local (re)investment				
	Community infrastructure (e.g., roads)				
	Community harvesting				
	Access to community resources (e.g., water, grazing lands)				
	Community cultural landmarks				
	Community recreational activities				
	Land tenure				
Equity Who benefits?					
	Women				
	Men				
	Girls				
	Boys				
	Old				
	Young				
	Poor				
	Rich				
	Illiterate				
	Literate				
Questions (Yes=positive impact)					
	Does the system or technology respond to problems and constraints identifed by villagers?				
	Were local people involved in all stages of the project planning / development?				
	Is the system or technology supported by the local power structure?				
	Is the system or technology compatible with current local practices, preferences, and wisdom?				
	Does the system or technology build on local practices and on existing capacity?				
	Is the system or technology supported by other factors (e.g., land tenure, macro policy)?				

Checklist Sustainability Concerning the Ecosystem

Project:____

Responsible:____

		negative	*positive*	*neutral*	*Remarks*
Biodiversity					
	Wildlife diversity				
	Rare or endangered species				
	Species abundance				
	Wild plant diversity				
	Crop diversity				
	Introduces exotic species? If yes, this could be a negative impact.				
Land					
	Wildlife habitat				
	Vegetation cover				
	Soil texturte				
	Nutrient recycling				
	Soil fertility				
	Soil structure				
	Soil or slope stability (e.g., erosion)				
Water					
	Water access				
	Water supply				
	Water quantity				
	Water quality				
	Drainage pattern				
Air					
	Air quality				
Resource use					
	Land use				
	Conservation of natural resources				
	Resource use				

7.3 Tools for Integrated Management of Process, Information & Quality

7.3.1 Flow Chart and Process Description

FC Holistic Development Management

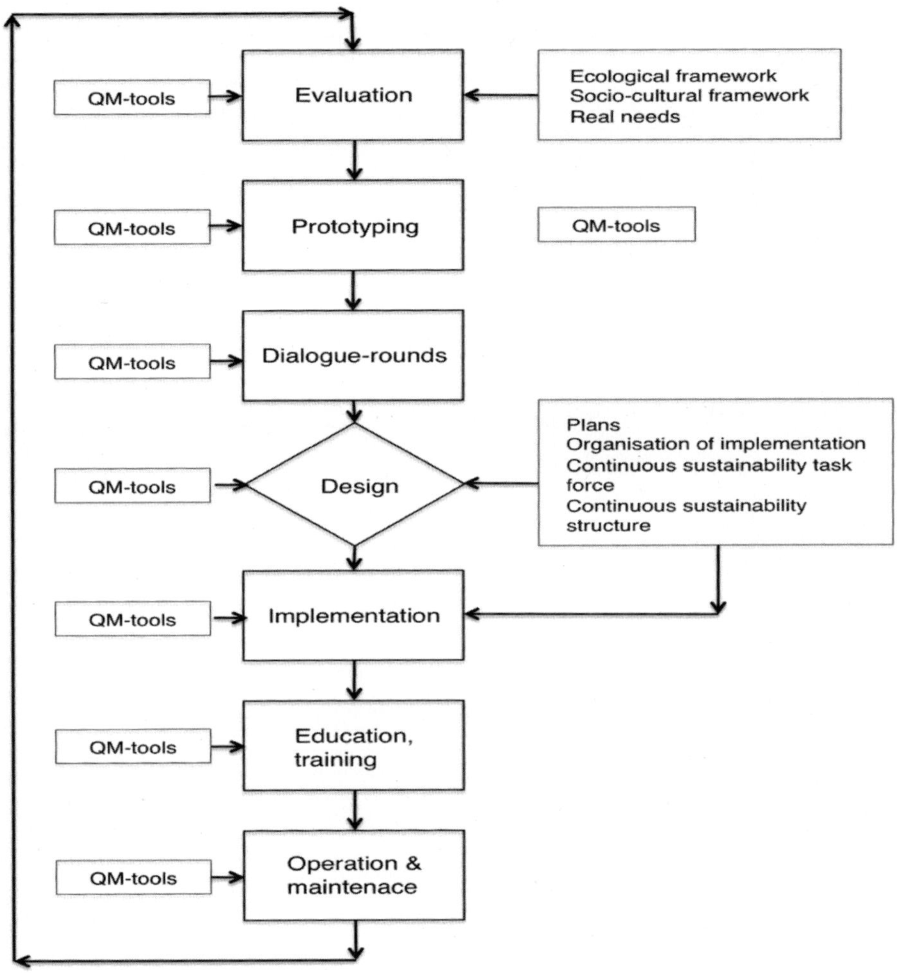

PD Holistic Quality Management

Input	Process	Result	*Stakeholders*
Request or enquiry	Start	Visit site	*Contracting body, interested contractor*
	↓		
Requirements ➤ CL Evaluation	Evaluation Step1	Known ecological and socio-cultural framework	Contracting body, contractor, interest groups,
	↓		
Real needs ➤ M1-3	Evaluation Step2	Known needs	*Significant stakeholders, coordinator*
	↓		
Expertise in ecological design	Prototyping	First proposals for plans/measures	*Designers*
	↓		
➤ M1-3	Dialogue-rounds	Accepted design	*Designers, stakeholders*
	↓		
Accepted plans and measures Professional literature	Design, Planning	Plans for implementation	*Designers, contracting body*
	↓		
Plans/designs	Implementation	Realized measures	*Supervision team, construction team*
	↓		
Plans, designs, operation guidelines/manuals	Training & education	Appropriate know-how	*Designers, supervisiors, operation team*
	↓		
	Appropriate structures	Sustainability	*Contracting body*
	↓		
Operation guidelines ➤ FO Operation manual	Operation & maintenance	Functioning measures	*Operation and maintenance teams*

7.3.2 Checklists

CL Sustainability of Human Ecosystem-Projects: Key Factors

Project: ____

Responsible____

	Yes	*No*	*ok*	*Remarks*
Clear definition of project goals				
Project is demand driven				
Project solves a real local problem				
Project brings social + economic benefit				
Socio-cultural framework understood?				
Ecological framework understood?				
Real local needs understood?				
Local decision-making understood?				
Prephase possible / done?				
Commitment of stakeholders				
Solutions functioning / effective?				
Long-term supporter/organisation available?				
Stable organization as client/main stakeholder				
Stability of country				
Stability of organizational unit				
Establishment of long-term unit as care-taker				
Harmony in project-group				
Harmony in local system?				
Fast decision making				
Enough time?				
Enough resources?				
Packaging and marketing effective				
Creating businesses that support long-term function of the project				
Financial participation of client and local stakeholders				
Participation of local stakeholders in project development?				
Infrastructure + resources for future activities?				
Action-research for participation?				

Remarks:

7.3.3 Forms

A very good tool of integrated process, information and quality management to follow the process of realization is a QM-questionnaire, filled by the respective participating members independently and confidently, thus allowing personal statements and insights, which might otherwise not be presented and available. The usual span of time of reporting is between 1 and 3 month. The coordinating person or institution by evaluating the various questionnaires simultaneously gets a good picture of the progress, the problems, possible solutions and the atmosphere of the project at this point in time.

A questionnaire is also good for getting feedback of workshops and conferences, best would be to have one filled before or latest at the start and another one at the end, summarizing the findings.

For each of the topics (project supervision, conference start, conference end) an example of questionnaire is made available in the following.

Depending on the mode of answering techniques (hand-writing or filling the word-document on the computer) the form has to be adjusted: for handwriting the boxes for the text have to expanded and then printed on paper.

All this forms have been tested and found very helpful in the development-projects at Valley View University.

QM-Questionaire for Project Supervision

Time-span of reporting, Project-Partner:

Aim: Exchange of information and experiences, to further the success of the project. Gives the assessment of the respective sub-project and from that viewpoint also on the whole project.

Time-span: Since last report. Delivery: first day of every second month.

Content/Style: The essence, concerning one´s own area of work and responsibility. In short notes. Your personal and subjective estimate is important and asked for. If there is really nothing to report in a specific field under question, simply leave it empty. The order of answering can be changed, the answer should cover the points in question however. You are free to add more topics. And if you have suggestions for improvement of this questionaire, you are welcome!

What has been achieved in the area of buildings / constructions / installations compared to the last report?

What has been achieved in the area of other measures compared to the last report?

What has been achieved in the area of research compared to the last report?

How about the state of project-schedule concerning buildings / construction / installations?

How about the state of project-schedule concerning implementation of measures?

How about the state of project-schedule concerning research?

What worked out especially bad in the area of measures / installations?

What worked out especially bad in the area of communication / information / cooperation?

What worked out especially bad in the area of research?

What worked out especially well in the area of measures / installations?

What worked out especially well in the area of communication / information / cooperation?

What worked out especially well in the area of research?

What should be done concerning implementation?

What should be done concerning cooperation / communication?

What should be done concerning research?

What was the most important since the last report?

Are there new ideas concerning new projects / possibilities?

Are there special wishes concerning the success of the project?

How about your satisfaction concerning the project? (bad-medium-well)

Important dates:

More?:

Name, place, date

ClimSys2010 QM-Questionaire (Starting phase): Your expectations and hopes

How? You can do the filling quick and with only few remarks and short sentences, just like the answers come to your mind. If you can´t find an answer to a question, simply skip it, at least for the moment. You can fill it anonymously if you want and your comments and suggestions are kept anonymously anyway, to allow for a free flow of communication. Your personal private subjective opinion or suggestion is most valuable and helpful, not so much official statements. You can add additional comments or suggestions as you like, also for the improvement of this questionaire. Thank you very much for your contribution!

Concerning ClimSys2010: What is most important for me?

What could ClimSys2010 contribute to my present situation?

What is the most important thing I do want to learn?

What is the most important thing I want to share?

Which suggestions do I have for the program?

Is there something else I want to remark?

Place and date:

More remarks

QM-Questionaire for the ClimSys2010 Workshop (Concluding phase): Your insights, conclusions and remarks

Why? You have been part of ClimSys2010 and we hope the time together was a joy for you.

Now concerning this joint experience, may we ask you a favour: Your sharing of comments and suggestions would be of greatest help and it is only you that can put in and contribute your insights, conclusions and remarks.

How? You can do the filling quick and with only few remarks and short sentences, just like the answers come to your mind. If you can´t find an answer to a question, simply skip it, at least for the moment. You can fill it anonymously if you want and your comments and suggestions are kept anonymously anyway, to allow for a free flow of communication. Your personal private subjective opinion or suggestion is most valuable and helpful, not so much official statements. You can add additional comments or suggestions as you like, also for the improvement of this questionaire. Thank you very much for your contribution!

Please tick the box under the best fitting comment for the following questions:

Question	**much**	**a little**	**no**
Did I enjoy the workshop?			
Did the workshop enlarge my kowledge about existing eco-projects, eco-networks, ecostudies and -trainings?			
Was the workshop helpful for enlarging your commitment to eco-projects, eco-networks, ecostudies and –trainings?			
Was the composition of the program (circle, workgroups, excursion, world-cafe, presentations, open market...) appealing?			
Was the map with some information materials helpful for me?			

Concerning the workshop: What was most important for me?

What was my high-light experience at the workshop?

What was the most important thing I learned?

What was the best at the workshop to my opinion?

What was the worst at the workshop to my opinion?

Which suggestions do I have for the next workshop concerning it´s content?

Which suggestions do I have for the next workshop concerning it´s organisation?

What else should be improved?

Which suggestions do I have for a next joint eco-project?

What do I think most important for the success of eco-projects/eco-studies?

If I would be the responsible leader for an eco-project: what would be my priorities?

If I would be the responsible leader for an eco-project: what would be my immediate next steps?

Is there something else I want to remark?

More remarks:

Place and date:

Fig. 7.1 Closing Circle of ClimSys2010 in front of the new Baobab-Center for ecological studies at VVU

7.3.4 Methods

7.3.4.1 Circle

Background

Many indigenous cultures use this method for communication, where in a round of people all without hierarchical concerns can express themselves without being interrupted by others.

Application

Good for the start and end of projects, workshops, seminars, training programs. Good for less than 40 people.

Method

All persons sit in one circle. One after the other speaks and expresses himself for as long as he likes and is appropriate for the purpose (normally some few minutes, a little longer for smaller groups, a little shorter for larger groups) (Fig. 7.1).

Fig. 7.2 World-Café round of ClimSys2010 at the new Baobab-Center at VVU

7.3.4.2 World-Cafe

Background

This method enables closer acquaintance with each other and fast exchange even immediately after first coming together.

Application

Good for first steps in workshops or seminars.

Method

Four to five persons sit around one table. One person as host writes down what emerges in the round, best on a flipchart sheet in big letters and short summarizing sentences with only few words and one line for one idea preferably. This is brainstorming time, not arguing pros and cons! After 15–20 min all except the host leave for other tables. The host explains to the newcomers the findings of the former round shortly, the next round starts for another 15–20 min. This should be repeated two or three times for one topic or step in the workshop. The findings can be presented to all by the hosts of the table (Fig. 7.2).

7.3.4.3 Appreciative Inquiry

Background

Concentrating on the positive only enables a supportive flow of thoughts and actions, whereas a list of failures and negative topics tends to hinder progress.

Application

Good for every eco-project.

Method

This is a four-step procedure (named: discovery, dream, design, destiny), concentrating on the positive side of the project or organization.

It starts with the evaluation of the existing positive (discovery), followed by the vision (dream), the vision assuming, that there are no limits in respect to resources (financial, personnel etc.). Grounded on the existing positive and the vision step three (design) explains the general form of realisation, be it finances, organisation, structure, procedure etc. The last step (destiny) goes even further in giving the first few next steps for bringing the vision nearer to realisation, naming the responsible persons/organisations, the necessary helpers and signs of completion. Appreciative Inquiry can be combined with world-café, where for each of the first three steps 2 or 3 world-café rounds are useful.

7.4 Tools for Nature Conservation and Vegetation Management

7.4.1 Flow Charts and Process Descriptions for Nature Conservation and Vegetation Management

Flow Chart (FC) Conservation of Natural and Semi-natural Habitats

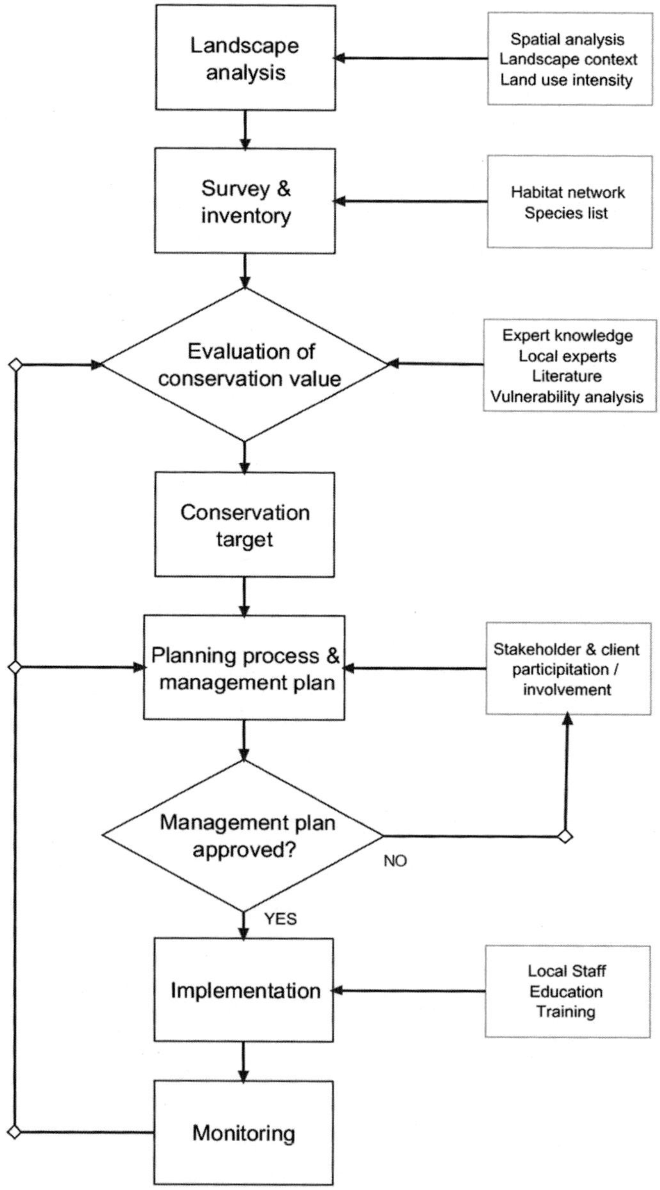

Process Description (PD) Conservation of semi-natural and natural habitats

Input	Process	Result	*Main contributing actors*
Request or enquiry	Start	Visit site	*Contracting body, interested contractor*
Maps Local knowledge Literature	Landscape analysis	Landscape evaluation	*Local experts, authorities* *stakeholders,*
Masterplan Local knowledge Literature	Survey & Inventory	Habitat and species evaluation; conservation target	*Local experts, authorities* *stakeholders,*
Evaluation results	Planning	Recommendations for management plan	*Local experts*
First recommendations, alternatives	Dialogue	Improved management plan	*Local decision makers, community*
Final recommendations	Management Plan	Accepted management plan	*Local decision makers, community*
Measures	Implementation	Implemented preservation areas & management	*Local decision makers, staff, community*
	Maintenance / Protection	Sustained preservation of habitats	*Staff*

Flow Chart Tree Planting Project

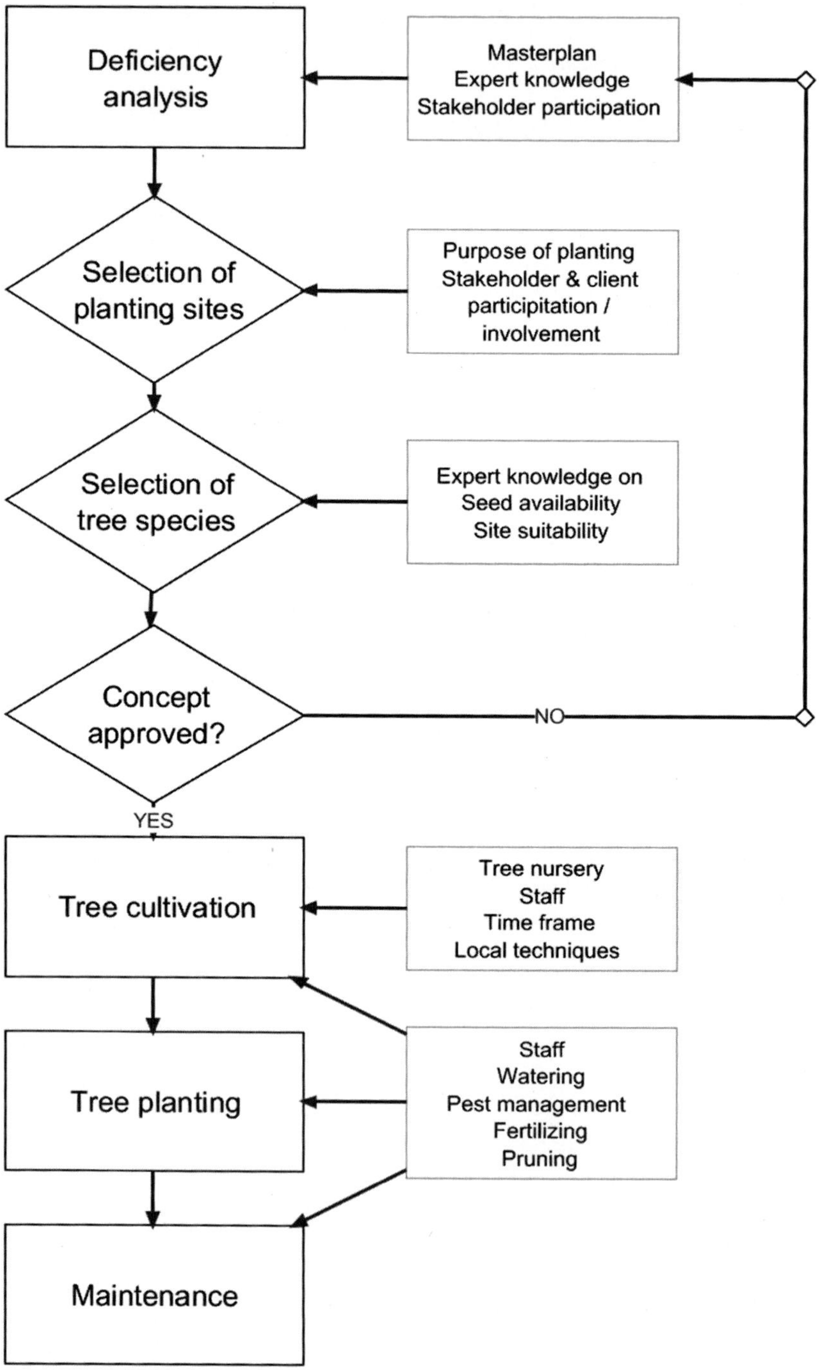

Process Description (PD) Tree planting project

Input	Process	Result	*Main contributing actors*
Request or enquiry Needs	Prephase	Aims & targets of plantings	*Contracting body, interested contractor*
	↓		
Masterplan Expert knowledge Stakeholder Involvement	Deficiency Analysis	Deficiency report	*Consultant, experts, stakeholders*
	↓		
Local knowledge Conservation areas Literature	Site Selection	Selected sites, Species suitability	*Contracting body, local experts*
	↓		
Seed & stock availability Site suitability	Species selection	List of chosen species	*Experts*
	↓		
List of chosen species Infrastructure (e.g. space/water) Local expertise & techniques ➢ CL Nursery requirements	Tree Cultivation	Cultivated trees	*Experts, staff*
	↓		
Cultivated trees	Tree Planting	Planted trees	*Staff, experts*
	↓		
Irrigation, fertilizer, pruning ➢ CL Vegetation maintenance	Maintenance	Sustained plantings	*Staff, experts*
	↓		

7.4.2 Checklists for Nature Conservation and Vegetation Management

CL Requirements for Tree Nursery

Project:____

Responsible:____

	Yes	*No*	*ok*	*Remarks*
Available space sufficient?				
Sufficient trained staff available?				
Sufficient work force available?				
Shaded areas required?				
Sufficient shaded areas availabe?				
Sufficient irrigation water available?				
Sufficient potting soil / substrate available?				
Pest management knowledge available?				
Budget sufficient?				

Remarks:

CL Planting and Maintenance of Tree Plantations

Project:____
Responsible:____

	Yes	*No*	*ok*	*Remarks*
Planting				
Workforce sufficient?				
Supervision available?				
Transport of seedlings to site secured?				
Tools adequate in type and numbers?				
Sufficient spacing of trees?				
Maintenance				
Sufficient irrigation water available?				
Water transport secured?				
Pipes, tanks, cans & tools available?				
Work force sufficient?				
Budget sufficient for 12 months irrigation?				
Fertilizer / manure available?				
Pest management (if necessary) secured?				
Regular pruning secured?				

Remarks:

Planting: The optimal time for planting trees in the tropics is early in the rainy season. This reduces the expenses needed for irrigation considerably. Also the planting is more laborious in the hardened soils during dry season. The spacing of trees is determined by the actual size of the grown trees.

Maintenance: Depending on the frequency of rains and the availability of water in the soil, a well-adapted irrigation scheme is mandatory to sustain the plantings. During dry periods a weekly scheme might be necessary. Irrigation should be secured for at least 12 months.

7.5 Tools for Mass Flow Management

7.5.1 Flow Chart and Process Description

FC Mass Flow Management

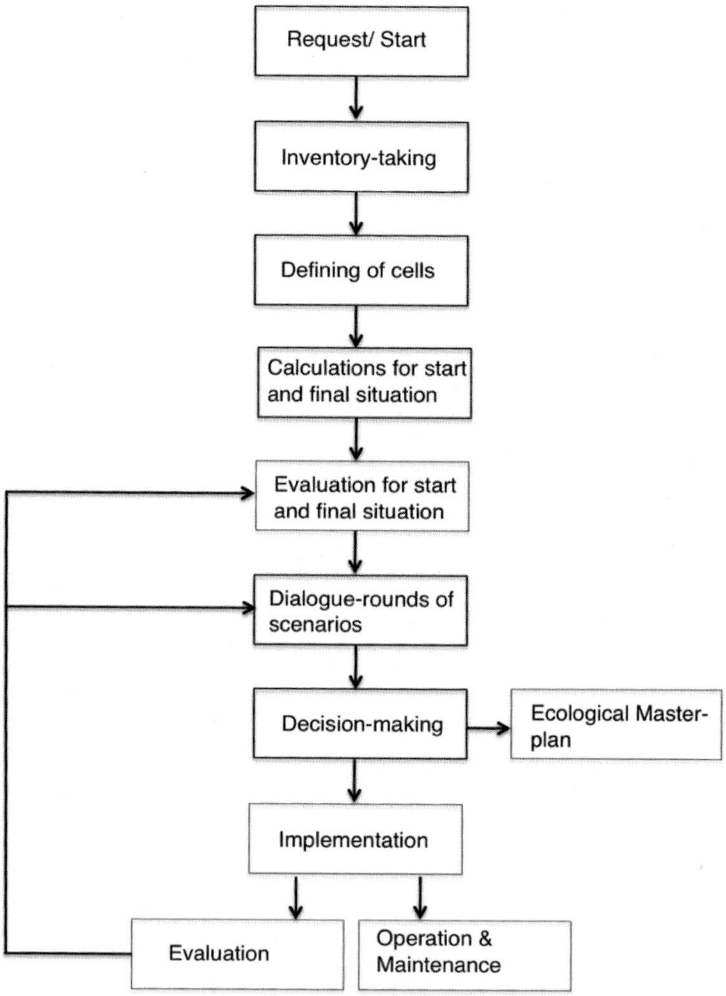

Process Description (PD) Mass Flow Management

Input	Process	Result	*Stakeholders*
Request or enquiry	Start	Agreement	*Contracting body*
Measurements Plan of installations	Inventory-taking	Mass-flow/ Installation	Internal & external experts
Characteristics of mass flow units	Defining of cells	Defined cells	*Experts*
Defined cells, result of measurements	Calculation for start & final stage	Results for start & final stage	*Experts*
Result on calculation	Evaluation for start & final stage	Conclusions, pros and cons	*Experts*
Summary of conclusions and pros and cons	Dialogue-rounds	Understanding of choice and preferences	*Experts, contractors, stakeholders, NGOs, moderator*
Summary of dialogue-rounds	Decision-Making	Decisions for further development modified master-plan	*Experts, contractors, stakeholders, moderator*
Masterplan, Plans, Money	Implementation	Implemented measures	*Construction companies, installation personnel*
Measures & Installations	Operation & Maintenance	Functional system evaluation	*Operation personnel*
Findings for QM	Evaluation engaging	Results, recommendation	*QM-responsible personnel*

7.5.2 Checklists for Mass Flow Management

CL Mass Flow Management: Inventory-taking for Waters

	Done	To be done	Not necessary	Remarks
water meter-reading income				
water meter-reading on campus				
water meter-reading buildings				
amount of shower-water				
amount of water for toilets				
amount of water for eating				
amount of drinking-water				
mass-flow for cafeteria				
mass-flow for dormitories				
roof-area collecting rainwater				
roof-area total				
surface-area collecting rainwater				
total surface area				
amount of water for gardening				
amount of water for agriculture				
amount of water for factory/ handcraft				
amount rainfall				
seasonal rainfall				
exaggeration				
amount of wastewater				
amount of sludge				
amount of drinks, sachet water etc.				
quality of pipeline				
quality of installations				
quality of personnel				
responsibilities of personnel				
funds for aggregation & maintenance				

Remarks:

7.6 Tools for Water Management

In this chapter, information related to water management is provided in the form of flow charts (FC), process descriptions (PD) and checklists (CL). The objective is to provide the reader with some tools, which should help him to work on the development of water infrastructures. This chapter does not yet encompass the whole spectrum of water issues but does focus on some aspects of water supply systems. The text contains not only information related to the planning of water supply systems in general, but also includes more specific information on rainwater systems. Future edition of this manual will enclose more detailed information on other key topics of water management such as ecological sanitation. For now, the reader is invited to explore the content of the excellent publications listed at the end of this chapter.

Basically the whole processes of developing a water supply system and a rainwater harvesting system is depicted in flow charts (FC) and in sheets entitled process descriptions (PD). With the PD sheets, it is only possible to represent the processes as linear processes, i.e. going from one step to the next one. However this is not always the case during projects. Sometimes new information will oblige the project partners to go back to a project step which has been previously undertaken. It might be for instance necessary to adapt the project objectives. In this manual, such mechanisms are only depicted in the FCs. This is the reason why the reader should always used the PDs together with the FCs.

The checklists contain information which will be useful for most projects. However the reader must keep in mind that these lists are not exhaustive and that adaptation might be necessary depending upon the project.

7.6.1 Water Supply System

FC Development of Water Supply System

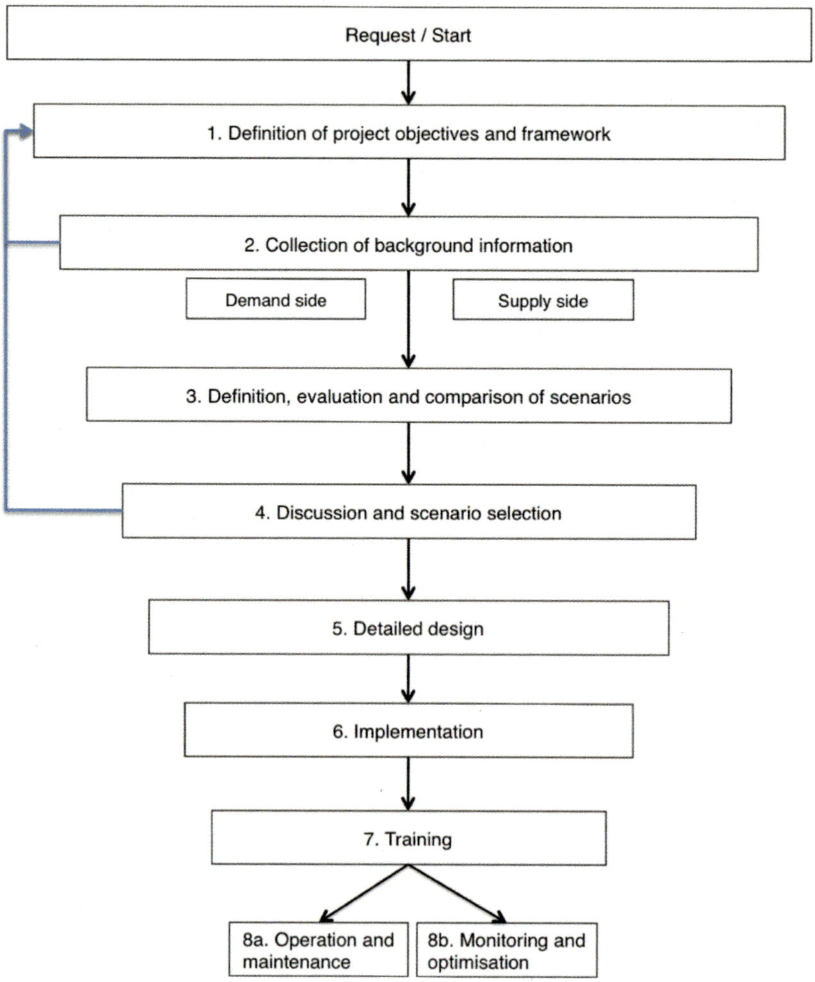

Comments:

The blue arrows represent the possibility of having to re-define or adapt the project objectives during the project development. This might be necessary after the collection of background information (Step 2). For instance, new information may reveal that some assumptions made during the definition of objectives (Step 1) were wrong. It might be also necessary when possible scenarios are discussed (Step 4). This would be necessary if none of the scenarios is feasible or enable to reach the objectives defined.

PD Implementation/further development of water supply system

Tools	Process	Result	*Stakeholders*
	Start	Project started	*Contracting body*
CL Project objectives and handling: important aspects to consider	1. Define project objectives and framework	Objectives and framework defined	*Contracting body, interested contractors, other relevant stakeholders*
CL Background information: water demand CL Background information: water resources	2. Collect background information (water demand and water resources)	Background information collected	*Contractors (planners) other relevant stakeholders*
CL Existing methods: water treatment and supply CL Evaluation of solutions: aspects to consider	3. Establish, evaluate and compare scenarios	List of scenarios established, evaluated and compared	*Contractors (planners) other relevant stakeholders*
CL Project objectives and handling: important aspects to consider	4. Select scenario	Scenario selected	*Contracting body, contractors (planners) other relevant stakeholders*
Professional literature	5. Design	Detailed design completed (e.g. plans)	*Contractors (designers) other relevant stakeholders*
Plans Professional literature	6. Implement	Scenario implemented (e.g. rainwater harvesting cistern constructed)	*Contractors (Supervision team, construction team), other relevant stakeholders*

Operation guidelines	7. Train responsible staff	Staff trained	*Contractors (Supervision team, operation and maintenance teams), other relevant stakeholders*
Operation guidelines	8. Operate and maintain Monitor and optimise	Facilities operated, maintained, monitored and optimised	*Operation and maintenance teams, other relevant stakeholders*

Comments:

CL Project Objectives and Handling: Important Aspects to Consider

Project: ____

Responsible:____

	YES?	NO?
Participation		
All relevant stakeholders are involved in the project.		
All stakeholders have the possibility to contribute to the project.		
The communication enables all stakeholders to be aware of the project development.		
Project objectives		
The objectives of the project are clearly defined.		
The project does not interfere with strategies defined by overruling authorities.		
The beneficiaries of the projects are clearly identified.		
The potential impacts of the projects (positive and negative) are recognized.		
Project handling		
The tasks necessary to achieve the project objectives are clearly defined.		
Adequate resources are allocated to each task.		
Responsibilities of project staff are clearly defined.		
The financial resources of the project enable are sufficient to reach the project objectives.		
All potential sources of funding are known.		
Local capacities and skills are used.		
The organisation of the community and its mechanisms are recognised. (e.g. administration procedures)		
Project time frame		
The project time frame is clearly defined.		
Project phases are clearly defined (e.g. by milestones).		

Comments:

CL Background Information: Water Demand

Project:____

Responsible:____

	YES?	*NO?*
Water demand		
The Community		
The needs of the community are identified.		
Scenarios of community development have been established.		
Culture, habits and beliefs with regard to water, waste water and hygiene are identified.		
The community ability and willingness to pay for water services are estimated.		
Sufficient information on the community have been collected (e.g. data, reports related to population, structure, economy and climate data).		
Water quantity		
An estimation of the current water demand is available.		
An estimation of the current water consumption is available.		
Scenarios for the future evolution of the water demand have been established.		
Large water consumers have been identified.		
Level of service desired has been defined (e.g. pressure requirements).		
Current issues with water supply are known.		
Water quality		
The water quality required is defined.		
The water quality parameters to consider are known.		
Legal requirements are known.		

Comments:

CL Background Information: Water Resources

Project:____

Responsible:____

	YES?	NO?
Water resources		
Type of resources		
Groundwater use has been considered.		
Surface water use has been considered.		
Rainwater use has been considered.		
Wastewater reuse has been considered.		
Quantitative aspects		
The possible intermittency of the source has been investigated.		
The yield of the water resource is known.		
Potential seasonal fluctuations of the yield are known.		
The minimum yield of the source has been compared with the community water demand.		
The recharge of the water source has been investigated.		
The quantity of water available is known or has been estimated.		
Legal, security or socio-political constraints to the use of the resource have been identified.		
Qualitative aspects		
The water quality of the water resource is known.		
Fluctuations of the water quality are considered / have been investigated.		
The vulnerability of the water resource to possible pollution has been evaluated.		
The potential for protection improvement of the water resources has been evaluated.		
Impacts of water use		
The impacts of water extraction/use on the environment, users and local population have been evaluated.		

Comments:

CL Methods Available: Water Treatment and Supply

Water resource: ____

Responsible:____

	Checked?
Water treatment and supply	
Pre-treatment	
Screening	
Roughing filtration	
Sedimentation (eventually with coagulation and flocculation)	
pH adjustment	
Pre-chlorination	
Aeration	
Treatment	
Chemical precipitation (e.g. for fluoride removal)	
Chemical oxidation	
Slow sand filtration	
Rapid sand filtration	
Membrane filtration (microfiltration, ultrafiltration, reverse osmosis)	
Clarification	
Activated carbon filtration/adsorption	
Disinfection (e.g. with UV-C or chlorine)	
Supply	
Gravity supply	
Pump driven supply	
General	
Centralised systems	
Decentralised systems	

Comments:

This checklist aims at providing the reader with a brief review of the most common water treatment and water supply features available. The water treatment and supply systems required would depend on the project (e.g. type of raw water). The systems must be designed by specialists.

CL Evaluation of Solutions: Aspects to Consider

Project:____
Responsible:______________________________________

	Checked?
Community aspects	
Impact of the solution on the community (e.g. feelings of ownership developed, responsibility).	
Acceptance by the community.	
Technical aspects	
Technical complexity of the solutions.	
Quality and lifespan of the equipments.	
Costs of the solutions.	
Benefits of the solutions.	
Technical requirements for operation and maintenance.	
Local availability of the material and technology required.	
Energy requirements.	
Staff requirements.	
Financial aspects	
Cost and benefits of the solutions.	
Financing of the solution.	
Tariff structure of water services for cost recovery.	
Life cycle assessment (investment, operation, maintenance, disposal)	
Institutional and legal aspects	
Capacity of local staff.	
Importance of participatory process.	

Comments:

7.6.2 *Rainwater Harvesting System*

This part provides some information related to the design of rainwater harvesting systems. Emphasis is put on the project development processes and key aspects to consider during the selection of rainwater harvesting systems. More detailed information can be found in textbooks entirely dedicated to rainwater harvesting systems (e.g. UNEP, 2002 or Thomas and Martinson, 2007).

FC Development of Rainwater Harvesting Systems

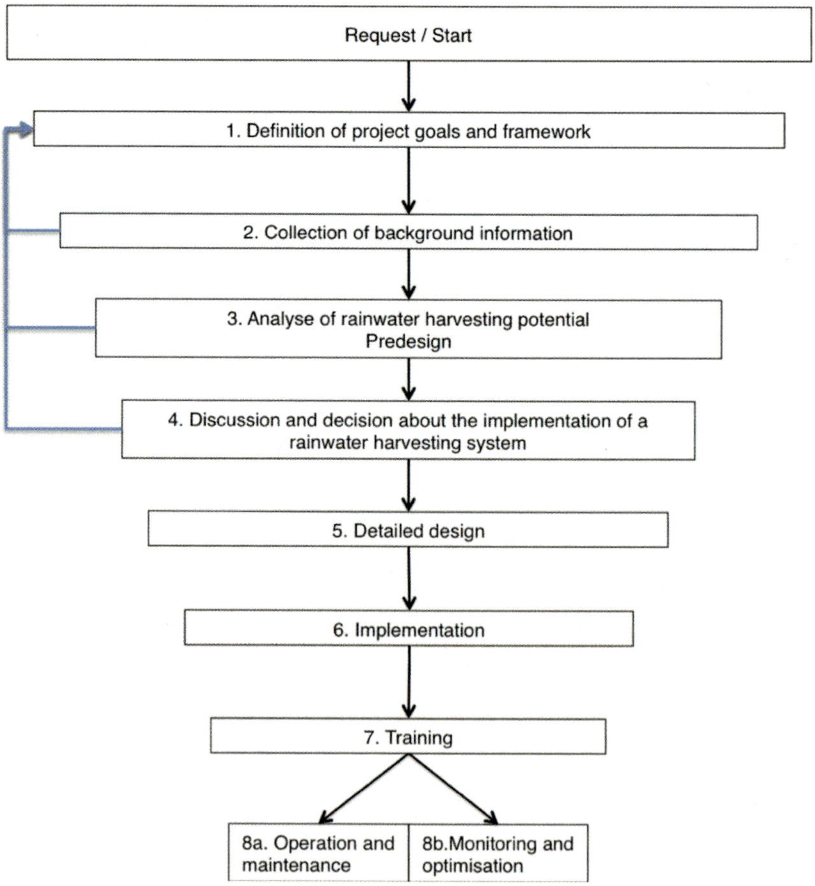

Comments:

PD Development of rainwater harvesting systems

Tools	Process	Result	*Stakeholders*
	Start	Project started	*Contracting body*
	1. Define project goals and framework	Project goals and framework defined	*Contracting body, interested contractors, other relevant stakeholders*
CL Background information: estimating rainwater harvesting potential	2. Collect background information	Background information collected	*Contractors (planners) other relevant stakeholders*
CL Background information: estimating rainwater harvesting potential	3. Analyse potential of rainwater harvesting + Predesign system(s) FC Predesign	Potential of rainwater harvesting estimated & System(s) pre-designed	*Contractors (planners) other relevant stakeholders*
CL Rainwater harvesting: aspects to consider	4. Decide which system to implement	Use of rainwater harvesting system discussed and system selected	*Contracting body, contractors (planners) other relevant stakeholders*
Professional literature	5. Design		*Contractors (designers) other relevant stakeholders*
Plans Professional literature	6. Implementation		*Contractors (Supervision team, construction team), other relevant stakeholders*

Operation guidelines	7. Train responsible staff	*Contractors (Supervision team, operation and maintenance teams), other relevant stakeholders*
	↓	
Operation guidelines	8. Operate and maintain Monitor and optimise	*Operation and maintenance teams, other relevant stakeholders*

Comments:

The implementation/development of rainwater harvesting system is here considered as one of several possible scenarios for the implementation/development of water supply systems. Other water supply alternatives should not be excluded.

FC Design of Rainwater Harvesting System

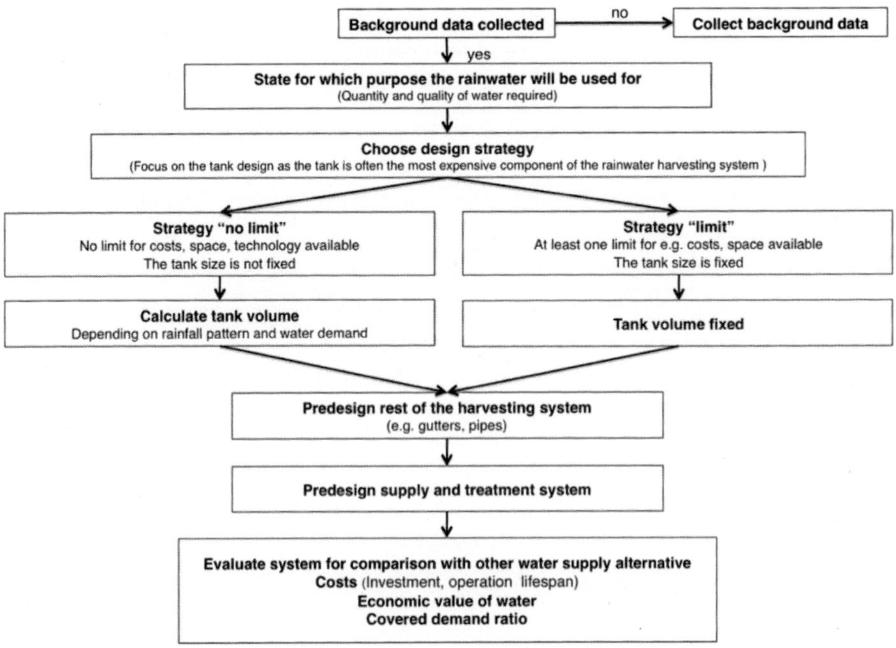

CL Background Information: Estimating the Rainwater Harvesting Potential

Project:____

Responsible: ____

	Checked?	*Comments*
Surface catchment		
Size of the surface catchment.	yes	400 m^2
Material of surface catchment (Runoff coefficient).	yes	Tiles (0,85)
Rainfall		
Annual Rainfall.	yes	700 mm
Annual quantity of rainwater that can be harvested.	yes	=(700/1000)*400*0,85=**238 m^3**
Rainfall distribution.	yes	2 rainy seasons (May-July; September-October)
Representative data available.	yes	Onsite measurements (daily measurements 2004 to 2009)
Water demand		
Average daily water demand.	yes	=20p*30l/p=0,6 m^3
Annual water demand.	yes	=0,6*365=219 m^3
Fluctuation of the water demand. (Constant? adaptive? seasonally-varied?)	yes	Constant
Potential use of the rainwater	yes	Toilet flushing, faucet
Potential of rainwater harvesting		
Is the annual quantity of rainwater which can be harvested higher than the annual water demand?	yes	238 m^3 > 219 m^3
Coverage ratio	yes	=238/219=108%

Comments:

CL Rainwater Harvesting: Aspects to Consider

Project:____

Responsible:____

	YES?	*NO?*
The future usages of the harvested rainwater are defined. (e.g. a sole source, the main source of water, wet season source, potable water source only)		
The distribution of the rainfalls throughout the year is known.		
Possible alternatives to rainwater use have been identified. (Other alternatives might be more practical or/and more economical).		
Rainwater use is a practical solution in that project.		
Rainwater harvesting and use is a solution accepted by the community.		
Rainwater use is an economic solution.		
Rainwater use is an affordable solution for the community.		
The community is willing to pay for rainwater use.		
The community has the capacity to manage the rainwater harvesting systems. (e.g. to reconnect disconnected gutter, to repair first-flush system, wash the filters)		
There is enough space available for the rainwater harvesting systems.		
The community has the capacity to design and implement rainwater harvesting systems on its own.		
The collection surface is large enough (e.g. ratio: roof size per inhabitant).		
There are any risks of contamination leading to a degradation of the water quality. (e.g. dust, insect, polluted air, intrusion of contaminants in the water, mosquito breeding, dissolution of materials)		
Monitoring of the water quality and of the quality of the installation (e.g. gutter well constructed).		

Comments:

7.6.3 Operation Guidelines for Water Treatment

The following pages were selected from the operation guidelines written for the operation and maintenance of the groundwater treatment plant at VVU. These guidelines (74 pages in total) were addressed to the teams in charge of water management at VVU and especially to the operator of the plant. The document aimed at helping the operator to operate and maintain the plant. It also provides information on the construction of the plant. The following is an excerpt, an abridged version of only some few pages giving an idea how such an operation guideline could look like.

Page 1: Front page
Page 2: Table of contents
Page 11: Description of the treatment plant
Page 16: Flow chart for the operation of the operation of the filtration units (1/2)
Page 17: Flow chart for the operation of the operation of the filtration units (2/2)
Page 22: Information on the plant maintenance
Page 23: Information on the monitoring of the plant
Page 73: Information on the construction of the plant
Page 55: Pictures showing how to operate the plant
Page 65: Gantt chart showing the main project steps

Groundwater treatment plant at Valley View University

–

Presentation and Operating instructions

Foreword

This document is addressed to VVU staff in charge of water supply management at VVU:

- Physical plant director
- Ecological manager
- Estate officer
- Plumbers
- Groundwater treatment plant operator(s)

The aim of the document is to enable the reader(s) to:

- understand why and how VVU groundwater is treated at VVU (Chapters 1 and 2 as well as Annex 1),
- provide all information necessary to operate and maintain the groundwater treatment plant (Chapter 3 and Annex 3),
- provide information related to the development of the groundwater treatment plant. Such information might be useful for future projects (Chapter 3.2, Annex 4 and 5).

With regard to the operating and maintenance instructions, which is the most important part of this document, the differences of educational backgrounds and working experiences between VVU staff members were considered. The instructions are given in two different ways: first as a text with some illustrations (Chapter 3), later as a sequence of pictures showing what must be done to operate and maintain the filtration units (Annex 3).

Author:

M.Sc. Richard Laurent, BOKU University

28/04/2010

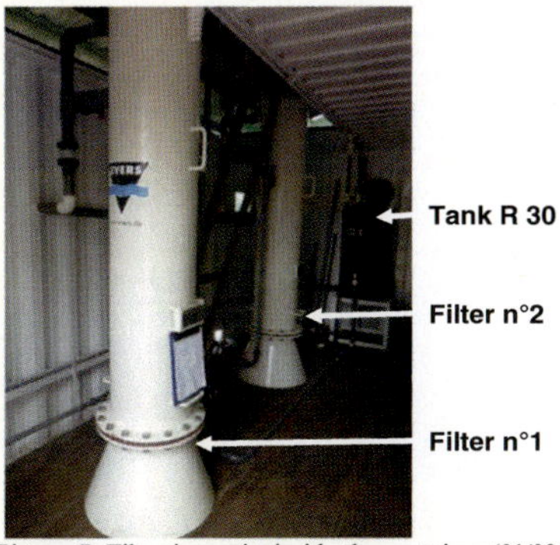

Picture 7: Filtration units inside the container (01/03/2010)

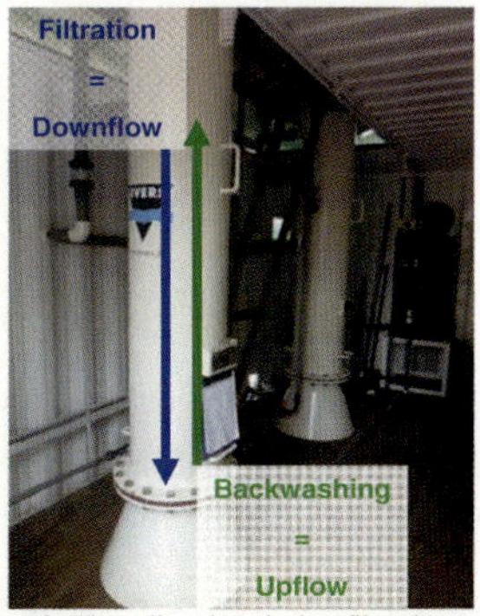

Picture 8: Flow direction during filtration and backwashing

Groundwater treatment at VVU is based on a down-flow filtration, i.e. water flows from the top to the bottom of the filter column. During the backwashing process, the water flows in the opposite direction, i.e. from the bottom to the top of the filtration unit. It is here worth mentioning that the cylindrical part of the filtration unit contains the 1.2 m (3.9") thick layer of filtration material, while its conic basement does contain only filtered water.

A schematic representation of the groundwater treatment plant (top view) is given below.

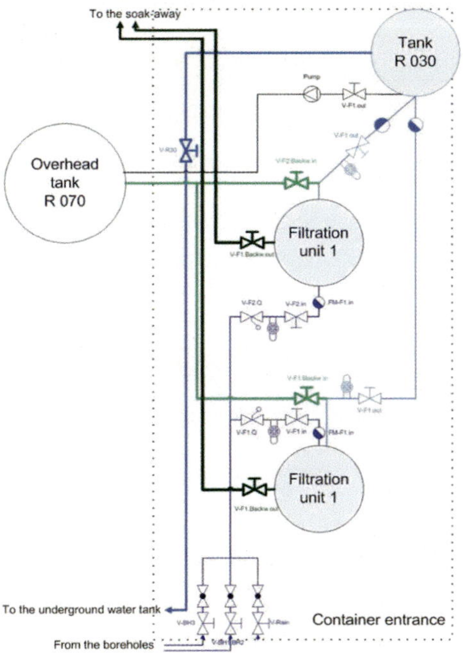

Picture 9: Schematic representation of the groundwater treatment plant (top view)

3.4.2 Operating instruction n°1: Starting the filtration units

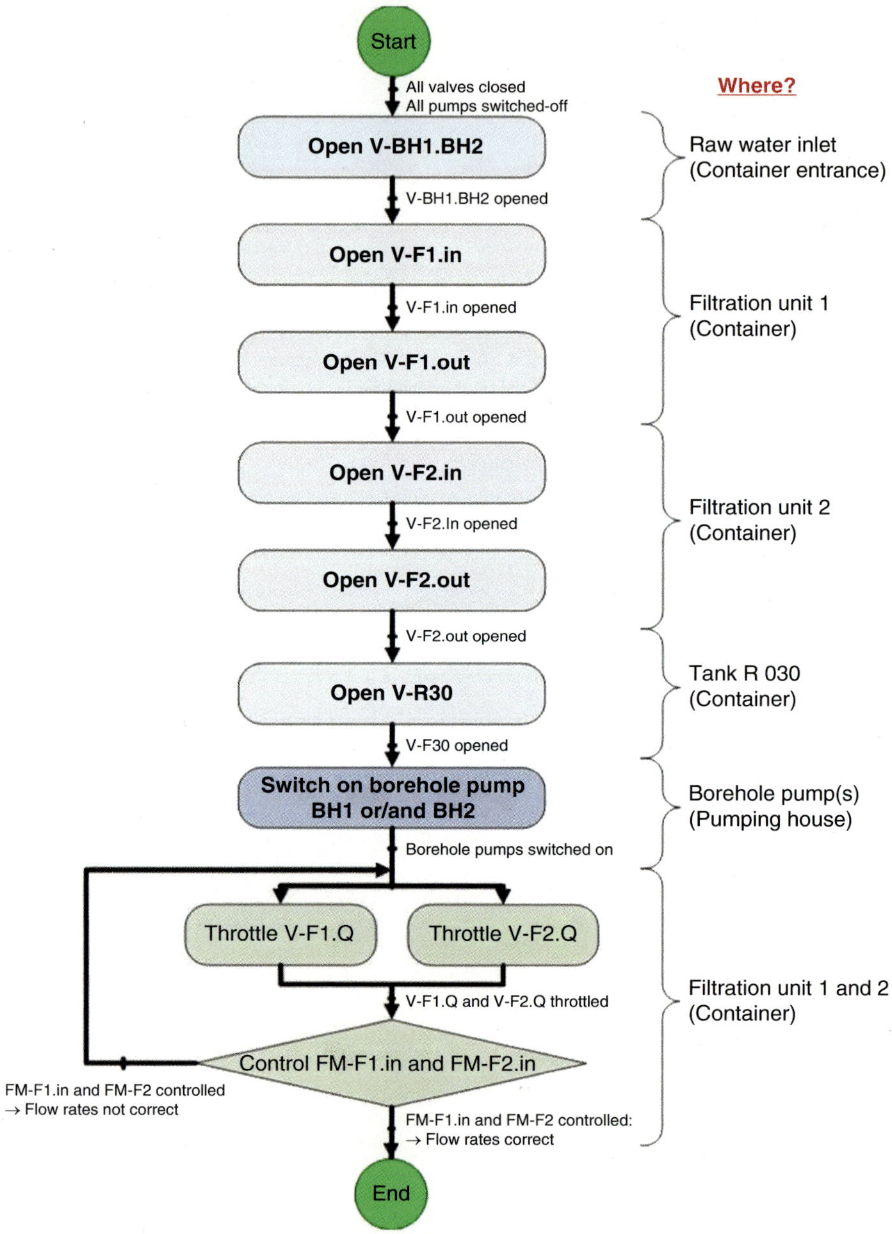

Picture 16: Operating instruction nº 1 – Starting the filtration unit

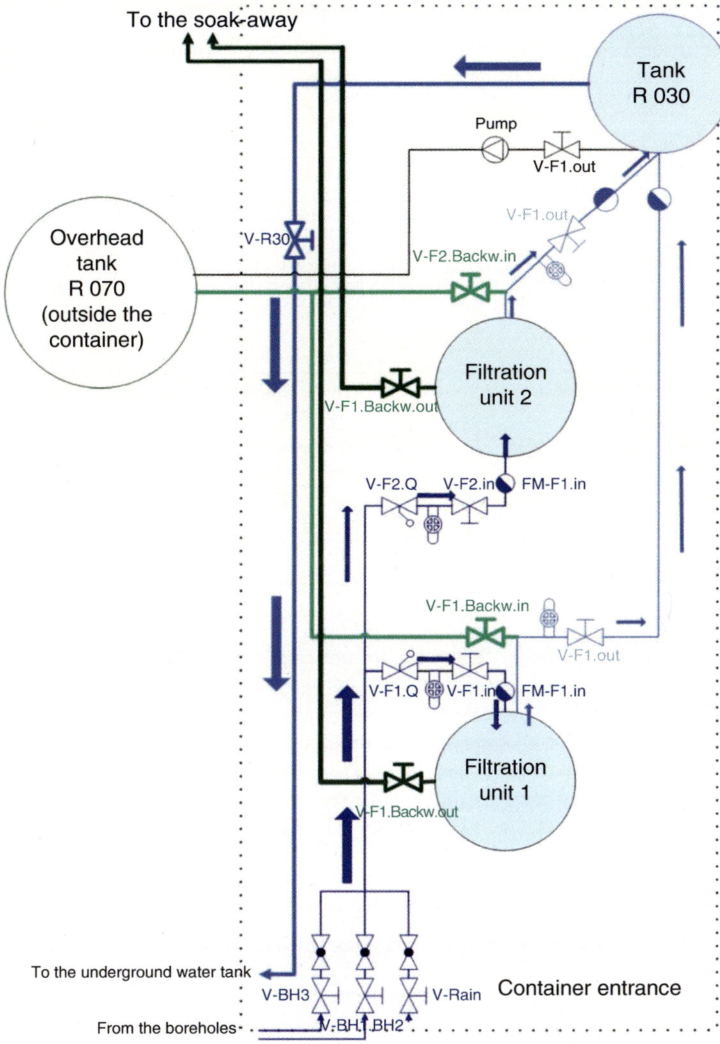

Picture 17: Schematic representation of the treatment plant - flow directions during the filtration process (top view)

3.4.7 Service and maintenance frequency

Table 7 summarizes the service and maintenance operations which need to be carried out for an optimal and long-term operation of the filtration units.

Table 7: Service and maintenance

What should be done	When
Backwashing	It depends on the raw water quality and the quantity of water treated. At least once a week.
Visual Inspection of the whole treatment plant	Once a week
Inspection of the injection points for flocculation and disinfection (if used)	Once a year
Replacement of the filter material	Every 6 – 8 years
Replacement of the flange gasket	At the same time as the filter material is being replaced
Cleaning of the whole treatment plant	It depends on the raw water quality and the quantity of water treated. Every 6-24 months

3.5 Trouble shooting

- **Overflow**: if the filtration units are overflowing, then control and compare the inlet flow rates of both filtration units (control flow meters labelled "*FM-F1.in*" and "*FM-F2.in*"). If the filtration flow rates are not equal, then adjust the valves labelled "*V-F1.Q*" and "*V-F2.Q*". If the filtration rates are correct, then the overflowing filtration units should be backwashed.
    - **Note**: For iron reduction, the filtration capacity of each unit is about 700-900 l/h (3500-4000 gallons/hour). The filtration units could be used at a higher flow rate, however, this would enable only to remove turbidity successfully and not iron (see also chapter 2.1, page 4)! Furthermore, the design of the treatment plant does not allow a higher flow rate. To increase the filtration flow rates, the diameter of the outlet pipes of each filtration unit should be increased (1" pipes are currently used).
- **Filtration material washed away during the backwashing process**: if accidentally a large amount of filtration material is washed away during the backwashing process, the lost material might be collected in the soak-away and reused. The only condition for a reuse is that the filtration material should still be clean.
- **No water is flowing**: *) check that the valves are opened (see Operating instructions page 16 to 21), *) check that the boreholes pump are switched on, *) check that the power supply is on.
- **Leakage**: if a leakage is observed (e.g. pipes, tanks), then stop the filtration units and repair the leakage.

3.6 Recording

The operator(s) should record every day data related to the operation of the groundwater treatment plant operation (see Picture 23). These data enable to record: *) the amount of groundwater extracted from the boreholes, *) the amount of water filtered and the backwashing frequency.

		Readings [m3]					Backwash		
Date	Time	Flow meter borehole 1	Flow meter borehole 2	Flow meter borehole 3	Flow meter filtration unit 1	Flow meter filtration unit 2	Backwash filtration unit 1	Backwash filtration unit 2	Comments
28.04.2010	08:00	225	123	30	172	172	No	No	No
29.04.2010	08:00	250	143	30	200	194	Yes	Yes	Some filtration material was washed away

Picture 23: Data to be recorded during the operation of the groundwater treatment plant

3.7 Sampling

Four taps were installed inside the container. These taps enable to collect water samples, which can then be used to evaluate the treatment performance of the filtration unit as well as the groundwater quality. The taps are located before and after the two filtration units.

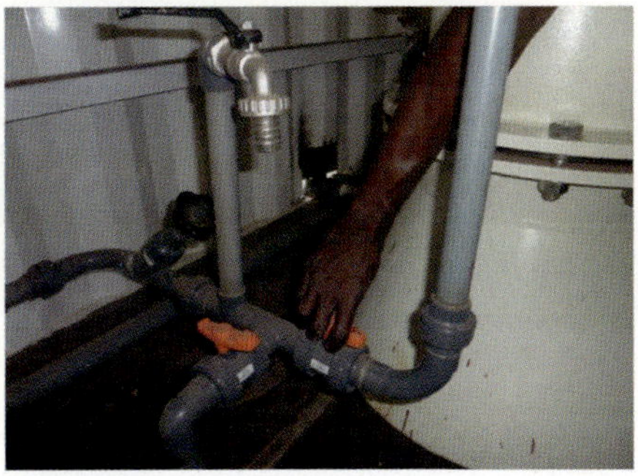

Picture 24: Taps for water sampling

Op. ins. 2-5. Open valve V-F1(or 2).Backw.out

Op. ins. 2-6. Open SLOWLY valve V-F1(or 2).Backw.in

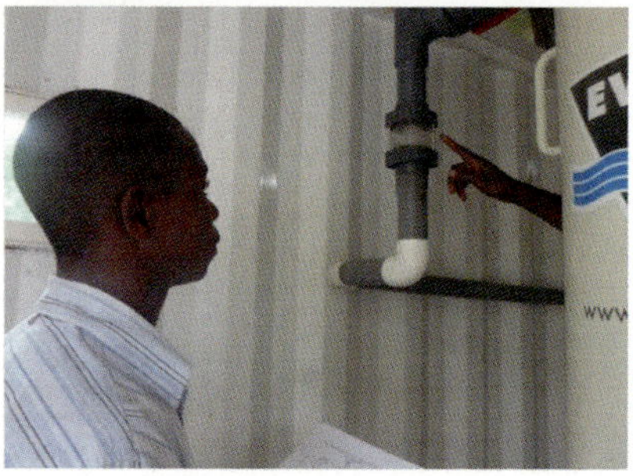

Op. ins. 2-7. Control transparent pipe

Op. ins. 2-8. If black filtration material is observed, reduce V-F1(or 2).Backw.out

Picture 83: Masons further working the soak-away (11/02/2010)

Picture 84: Carpenter and his team installing the mosquito net (11/02/2010)

Picture 85: Plumber installing the pipes connecting the filtrations units to the overhead tank (11/02/2010)

Picture 86: Plumber connecting the overhead tank (11/02/2010)

Picture 87: Painters starting to paint the container (12/02/2010)

Picture 88: First white coat of paint applied on the outside walls of the container (12/02/2010)

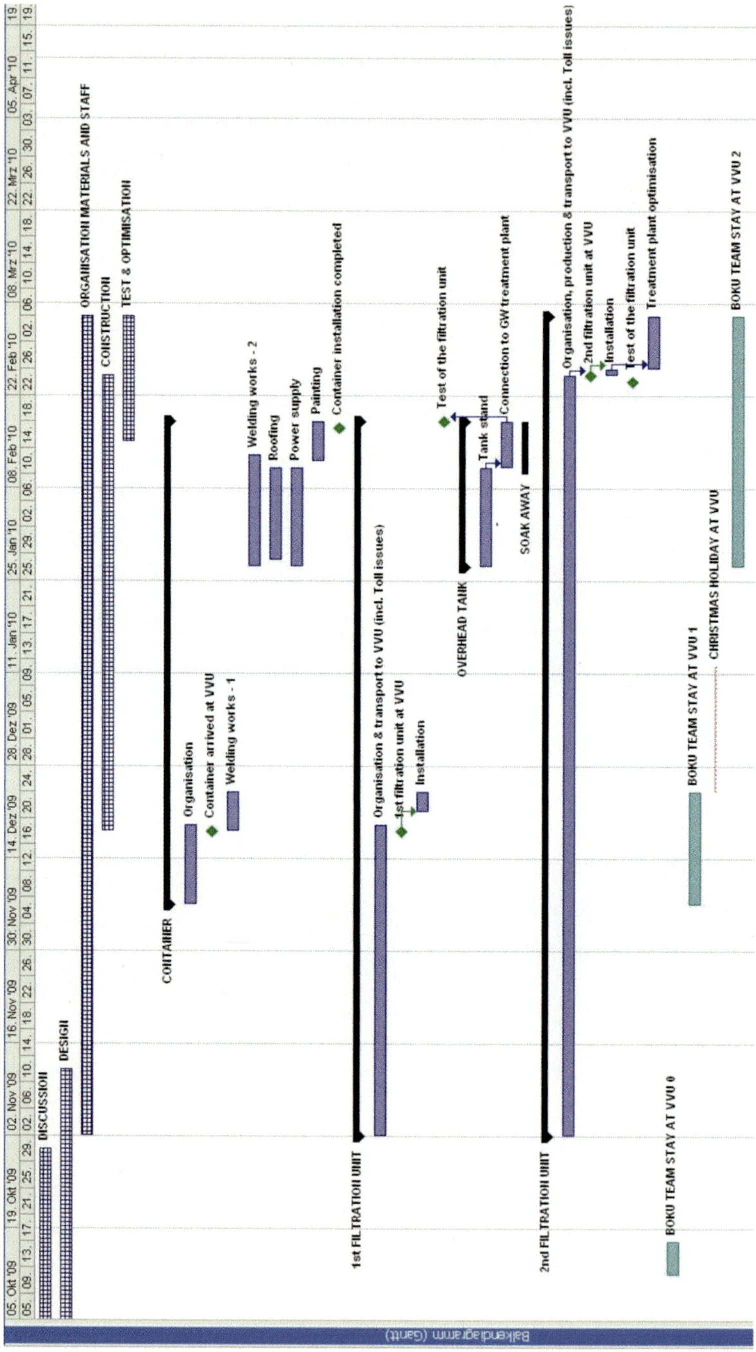
DISCUSSION
DESIGN
ORGANISATION MATERIALS AND STAFF
CONSTRUCTION
TEST & OPTIMISATION
CONTAINER
Organisation
Container arrived at VVU
Welding works - 1
Welding works - 2
Roofing
Power supply
Painting
Container installation completed
1st FILTRATION UNIT
Organisation & transport to VVU (incl. Toll issues)
1st filtration unit at VVU
Installation
Test of the filtration unit
OVERHEAD TANK
Tank stand
Connection to GW treatment plant
SOAK AWAY
2nd FILTRATION UNIT
Organisation, production & transport to VVU (incl. Toll issues)
2nd filtration unit at VVU
Installation
Test of the filtration unit
Treatment plant optimisation
BOKU TEAM STAY AT VVU 0
BOKU TEAM STAY AT VVU 1
CHRISTMAS HOLIDAY AT VVU
BOKU TEAM STAY AT VVU 2
Balkendiagramm (Gantt)

7.7 Tools for Ecological Building

7.7.1 Flow Chart and Process Description

Flow Chart Ecological Masterplanning

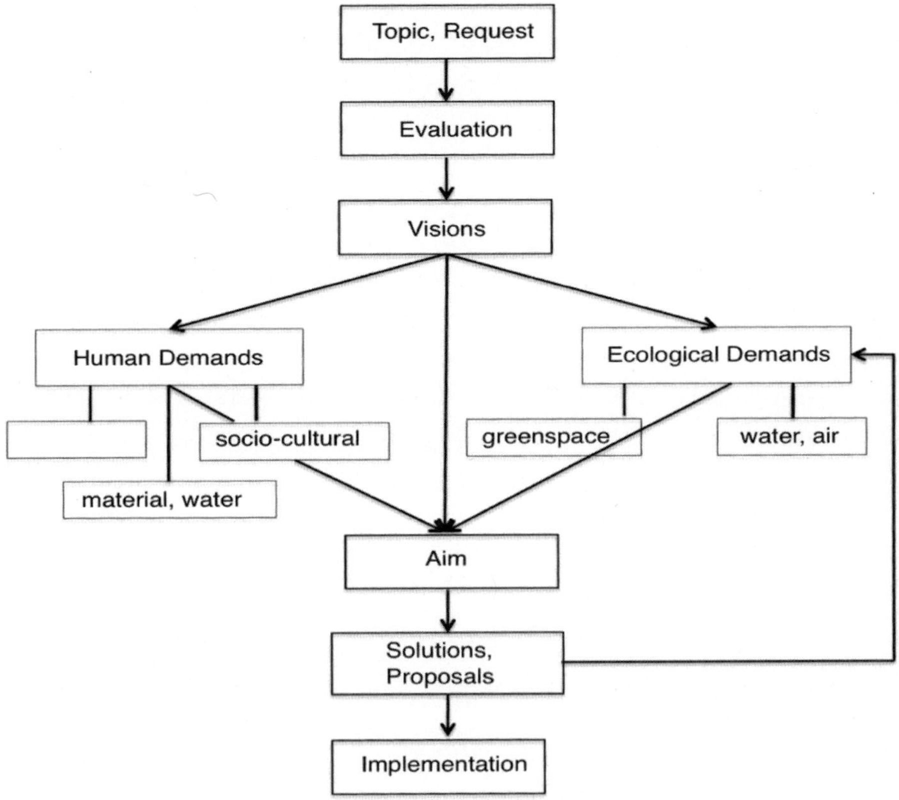

Annex (More Materials)

Posters

Posters BMU-Project Climate Friendly VVU

G. Geller and D. Glücklich (eds.), *Sustainable Rural and Urban Ecosystems: Design, Implementation and Operation*, DOI 10.1007/978-3-642-28261-4,

Poster: Integrative Showcase Baobab Centre at Valley View University

CLIMATE CHANGE
INITIATIVE PROJECT
2009/2010

Integrative Showcase Baobab Centre for Ecological Studies

Centre of Ecological Studies
Office for Sustainable Design, Buchholz

Vegetation and Green Space
Justus-Liebig-University Gießen

Rainwater Storage
Ecological Engineering Society, Augsburg

Water treatment
University of Natural Resources and Applied Life Sciences, Vienna

Structural Design and Supervision
Broberg Consult Ltd., Accra

Ecological Engineering Society, IÖV
Valley View University, VVU
Financed by
Federal Ministry for the Environment, Nature Conservation and Nuclear Safety

The holistic approach of the Baobab Centre combines various conceptual fields such as architecture, energy, water, green space and thus allows the building to perform in a climate friendly way.

Rainwater storage under paved footpath

Installation of photovoltaic panels on roof structure

Landscaping Plan, Scale 1:500

Water targets

- Minimized water consumption
- Closed water loops, recycling and harvesting of waste water
- Water quality management

Water methods

- Rainwater harvesting from roofs and collected in underground storage beneath a paved footpath
- Rain water re-use in the Baobab Centre for sanitary facilities and for further use on campus
- Water saving appliances in the Baobab Centre and Sanitary block such as press taps and 3l flush toilets
- Treatment of waste water with 3 chamber septic tank and sandfilter
- Re-use of treated waste water for irrigation in green space
- Filtering of water and further use as drinking water

Energy targets

- Minimized power consumption
- Generation of power
- Comfortable microclimate

Energy methods

Passive:

- Orientation to wind and sun
- Natural ventilation through formblocks, louver windows and corridors
- Shading verandah and roof overhang
- Bright surfaces for daylight reflection and low heat absorption
- Vegetation in open space cooling air through transpiration

Active:

- Photovoltaic generating power (approx. 15000kWh/a) for local use
- Overproduction is fed into the national power grid

The Baobab Centre produces more power than it needs for consumption. Therefore it can be called "climate-neutral".

Green space targets

- Improvement of microclimate
- Educational purpose

Green space methods

Microclimate:

- Focal point of the design: the large remnant Baobab tree
- Generous areas of green space and a large number of smaller trees around the building provide a pleasant microclimate. The students will sit, study and discuss underneath the shading and cooling trees

Education:

- Mainly indigenous species were planted. Most of them are medicinal plants or have other traditional uses
- The plantings can thus serve as living study material in ecology courses to raise awareness for local natural resources and biodiversity issues
- The Baobab Centre is surrounded by native plants also symbolizes the entire VVU campus, which is surrounded by precious remnant areas of native savannah bushland (in the remote part of the property)
- The green space is thus also a reminder to preserve this natural heritage at VVU

Poster: Baobab Center for Ecological Studies at Valley View University

CLIMATE CHANGE INITIATIVE PROJECT 2009/2010

Baobab Centre for Ecological Studies

A climate-neutral faculty building

Design and Implementation
Prof. Dr.-Ing. Detlef Glücklich
Office for Sustainable Design, Buchholz, Germany

Detailed Design
Dipl.-Ing. Antje Holdefleiss M.Sc.
Prof. Dr.-Ing. Detlef Glücklich
Office for Sustainable Design, Buchholz, Germany

Structural Design
Dipl.-Ing. Christiane Bergmann
Broberg Consult Ltd., Accra, Ghana

Execution
Mr. Emmanuel Kwandahor
Physical Plant, Valley View University, Accra, Ghana

Supervision of works
Dipl.-Ing. Christiane Bergmann
Broberg Consult Ltd., Accra
Dipl.-Ing Antje Holdefleiss M.Sc.
Office for Sustainable Design, Buchholz, Germany

Vegetation and Green Space
Dr. agr. Dietmar Simmering
Prof. Dr. Dr. Annette Otte
Justus-Liebig-University Gießen, Germany

Photovoltaik
Mr. Henk Vermeer
Energiebau Sunergy Ghana Ltd., Peduase, Ghana

Ecological Engineering Society, IÖV
Valley View University, VVU
Financed by
Federal Ministry for the Environment, Nature Conservation and Nuclear Safety, Germany

Bundesministerium für Umwelt, Naturschutz und Reaktorsicherheit

The Baobab Centre for Ecological Studies was designed to be a centre for the exchange of information and knowledge in the field of environmental sciences based at Valley View University (VVU). In the past 10 years VVU has developed the first ecological university campus in Africa. The faculty for Ecological Studies will guarantee an ecological education and at the same time connect students and teachers to the ecosystems on campus. It will enable further development and maintenance of the concepts and installations in place as well as the distribution of information beyond the borders of the campus.

The new faculty building offers 732m² seminar/office area and 525m² verandah area accomodated on three storeys. An extension can be added at a later stage by copying the design of the small cube/office as shown in Ground Floor Plan below.
Open classroom seminars can be held in the courtyard.

The architectural design takes into account the specific factors of its location (genius loci). The roof and verandah provide shade for the façades most exposed to the sun. Natural ventilation is enhanced by permeable walls with ventilation blocks and louver windows. Each costum-made ventilation block incorporates a frame with mosquito netting.
Most building materials were locally produced and long transport routes as well as material wastage could be avoided to a great extent. Broken tiles on the verandahs give an example of successful recycling methods used for this building.

A water saving sanitary concept can be found in the sanitary block provided behind the Baobab Centre. Urine is collected and used as natural fertilizer for VVU agriculture.
The treated waste water is re-used for irrigation. The rain water is saved in a large underground storage of 300m³ and can be used when needed.
Solar panels on the roof not only cover the demand of electricity, they produce additional power which is fed into the national power grid.

view from Baobab tree into conrtyard

Verandahs and roof provide shape

Ventilation blocks allow natural ventilation

Marterial recycling of broken tiles

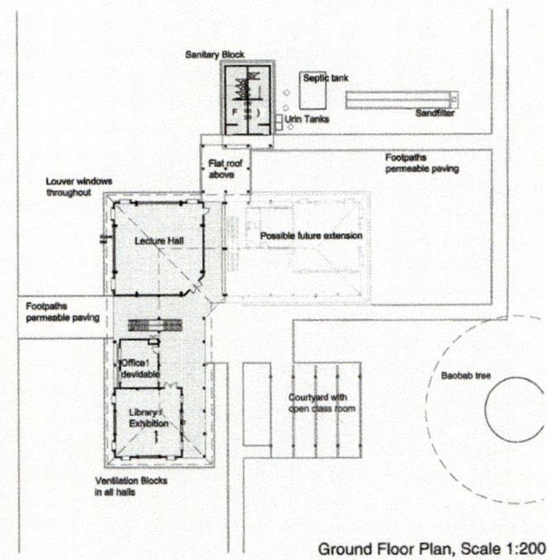

Ground Floor Plan, Scale 1:200

Poster: Tree Planting at Valley View University

CLIMATE CHANGE
INITIATIVE PROJECT
2009/2010

Vegetation and Green Space
Tree Planting

Design and Implementation
Dr. agr. Dietmar Simmering
Prof. Dr. Dr. Annette Otte
Division of Landscape Ecology and Landscape Planning
Interdisciplinary Research Centre (IFZ)
Justus-Liebig-University, Giessen, Germany

JUSTUS-LIEBIG-UNIVERSITÄT GIESSEN

Execution
Solomon Addai,
Agricultural Director
Physical Plant Department,
Valley View University

Consultancy and Tree Propagation
Prof. Dr. Joseph K. Osei, Provost
Valley View University, Techiman Campus

Ecological Engineering Society, IÖV
Valley View University, VVU

Financed by
Federal Ministry for the Environment, Nature Conservation and Nuclear Safety, Germany

Aims and concept

Global warming is mostly triggered by the so called greenhouse gases, like, e.g., carbon dioxide. Plants turn carbon dioxide and water into oxygen and sugar. They thus remove carbon dioxide temporarily from the atmosphere. Many scientists claim that due to this carbon storage effect, reforestation efforts in the tropics can help to mitigate climate change.

However, planting trees helps also in a more direct way to cope with the ever-warmer and drier weather, which we have to expect for the Greater Accra region. Mature trees create shade and decrease solar radiation and wind speed. This leads to an increase in humidity and an overall cooling effect for the microclimate - for the benefit of the people and the environment. Trees are also an important resource and habitat for many wildlife species that deserve our appreciation.

Measures

- **Fruit Tree Orchards** In addition to the already existing farmland (12.7 ha), another 6.1 ha have been planted with approx. 1420 fruit trees (oranges, mangoes, avocadoes etc). Water demanding trees like oranges will be irrigated with the collected grey water. Water pipes were appropriately installed. The harvest will be delivered to the cafeteria and sold to students and the local market.

- **Arboretum / Groves** The existing large Arboretum (1.9 ha) at the bakery has been planted with a considerable number of shade trees to become a place to rest, meditate and reflect underneath large trees. Other areas (2.1 ha) on the campus were, wherever suitable, also planted with a collection of different trees species. Eventually, these areas will also become shady groves and improve the microclimate.

- **Tree Islands** Within the more remote parts of VVU property, some areas have been repeatedly burned and overgrazed by unauthorized intruders over the years. Tree growth had been prevented. These areas (ca. 2 ha), have now been planted with suitable trees adapted to the dry environment. The idea is to foster the establishment of the local dry forest and to support wildlife and biodiversity.

- **Perimeter planting** The entire perimeter of VVU property was planted with a line of approx. 400 shade trees. Once grown, these trees will make VVU a visual landmark. Of course, these trees also improve the local microclimate and serve thus as a positive example of sustainable development for neighboring communities.

- **Avenues** The benefits of tree planting include of course also ornamental beautification purposes. Along the main roads 300 Veitchia palms were planted to support the landscaping design. Taller avenue trees were planted wherever suitable.

initial planting of a future grove

Example of a mature tree grove (Legon Univ. Arboretum)

Nursing seedings at VVU

Transport for planting

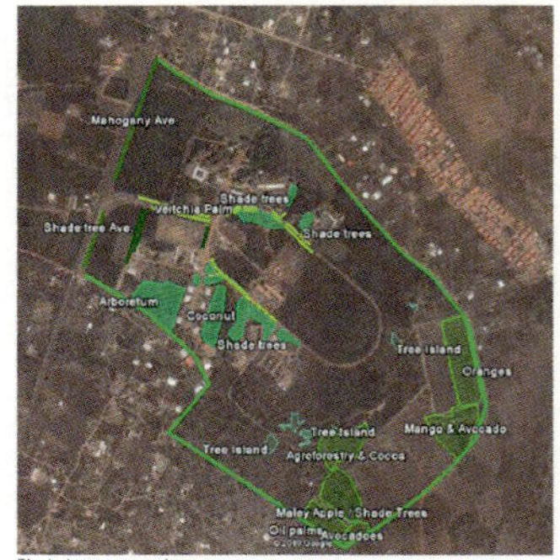

Planted areas - overview

Repotting Millettia seedlings to produce strong and healthy plants

Millettia - a beautiful indigenous shade tree

List of tree planted at VVU in 2009/2010. the total no. is 3,375

Number	English / Local name	Scientific name	Fruit Trees / Agroforestry	Groves / Arboretum	Tree Islands	Perimeter	Avenue
620	Sweet Oranges	*Citrus sinensis*	x				
300	Mango	*Mangifera indica*	x				
270	Dry Zone Mahogany	*Khaya senegalensis*	x	x	x	x	x
200	Cocoa	*Theobroma cacao*	x				
190	Ofram	*Terminalia superba*	x	x	x	x	
180	Avocado	*Persea americana*	x				
175	West African Oil Palm	*Elaeis guineensis*	x				
100	Coconut palm	*Cocos nucifera*	x				
65	Maley apple	*Syzygium malaccense*	x				
50	Sandbox Tree	*Hura crepitans*	x	x	x		
15	Guava	*Psidium guajava*	x				
160	Millettia / Sante	*Millettia thonningii*		x	x		x
160	Dahoma	*Piptadeniastrum africanum*		x	x		
140	Cedrela / Spanish Cedar	*Cedrela odorata*		x		x	
65	Lebbeck	*Albizia lebbeck*		x	x		
50	Rain Tree	*Samanea saman*		x	x	x	
35	Dawadawa	*Parkia biglobosa*		x	x		
25	Akee apple	*Blighia sapida*		x			
20	Kusia	*Nauclea diderrichii*		x	x		
20	Flamboyant	*Delonix regia*		x			
20	Prekese	*Tetrapleura tetraptera*		x	x		
10	Ordeal tree	*Erythrophleum suaveolens*		x	x		
10	Jack Fruit	*Artocarpus heterophyllus*		x			
8	Otie	*Pycnanthus angolensis*		x	x		
8	Edinam	*Entandrophragma angolense*					
7	Ashanti plum	*Spondias sp.*		x	x		
10	Voacanga	*Voacanga africana*		x			
7	Copper Pod	*Peltophorum pterocarpus*		x			
5	Catapa	*Terminalia catappa*		x			
5	Bread Fruit	*Artocarpus altilis*		x			
300	Veitchia palms	*Veitchia sp.*					x
95	Onyina/ Silk Cotton Tree	*Ceiba pentandra*			x	x	
40	Baobab	*Adansonia digitata*			x	x	
10	Wawa	*Triplochiton scleroxylon*			x		

Poster: Nature Conservation at Valley View University

CLIMATE CHANGE
INITIATIVE PROJECT
2009/2010

Vegetation and Green Space

Nature Conservation

Design and Implementation
Dr. agr. Dietmar Simmering
Prof. Dr. Dr. Annette Otte
Division of Landscape Ecology and Landscape Planning
Interdisciplinary Research Centre (IFZ)
Justus-Liebig-University, Giessen, Germany

JUSTUS-LIEBIG-UNIVERSITÄT GIESSEN

Execution
Solomon Addai, Agricultural Director
Physical Plant Department, Valley View University

Support Plant Identification
Mr. Daniel K. Abbiw, Accra
Mr. R.S. Amponsah, Accra

Ecological Engineering Society, IÖV
Valley View University, VVU

Financed by
Federal Ministry for the Environment, Nature Conservation and Nuclear Safety, Germany

Background and aims

- VVU lies within the dry forest zone of the Accra plains; an area strongly affected by urban sprawl that does not leave much space for natural habitats. The property covers 120 ha. Considerable portions of it are yet undeveloped. This land is covered with either savannah shrubs and trees or exotic, invasive species.
- Despite the ecological goals, the further development of the university requires building activities that will claim large portions of the yet-undeveloped land. To assess the ecological value of the remaining shrubland with respect to the preservation of local natural resources and biodiversity, we carried out a floristic survey.
- Our aim was to determine priority areas for nature conservation on campus.

Precious savannah shrubland and threats

- The floristic survey revealed a large number of 97 indigenous species typical of the natural wooded savannah. Among them are 26 shrub species, 23 climbers (lianas) and 26 tree species. Many of the recorded tree species on campus were only found in small numbers and as juvenile or coppiced individuals.
- Due to unauthorized firewood chopping by rural dwellers, occasional fires set by hunters and farmers, and a severe overgrazing of open areas by cattle, fully grown trees do almost not occur. 27 species found on campus are listed as 'endangered medicinal plants' on a national scale. Most of them are rare due to habitat destruction and overexploitation for medicinal purposes.

Measures

- **Preservation areas** The results show that it is essential to preserve larger portions of the natural shrubland. Designated areas (18 ha) have been suggested for preservation. Conflicts with future development will be negligible.
- **Fencing** To prevent future fires, as well as unauthorized hunting, cattle-grazing and firewood-gathering a fencing of the property will be inevitable. It is also needed to protect the planted trees.
- **Security** After a severe bushfire had affected the newly planted oil palm plantation in September 09, the security staff patrols the remote parts of the property more often.
- **Giant old trees** One old Baobab tree just outside VVU and one old Kapok (Silk Cotton) tree will be preserved as natural landmarks and symbols of the sustainable development of VVU.

Location of proposed Preservation areas

Eulophia cristata - a savannah orchid

Ruspolia hypocrateriformis - a savanah shrub

Giant Silk Cotton /kapok tree at VVU border

Lannea barteri - a rare savanah tree

Gardenia temifolia - a savanah shrub

Scadoxus multiflorus - a Fire Ball Lilly

Giant Baobab tree at VVU border

Clausena anisata - Mosquito plant

Ximenia americana - Wild Line

All pictures were taken at VVU - Examples of indigenous savannah plants

Poster: Water Treatment at Valley View University

CLIMATE CHANGE INITIATIVE PROJECT 2009/2010

Water treatment

Testing, treatment and monitoring

Testing, Design and Implementation

Dipl.-Ing. Laurent Richard M.Sc.

Dipl.-Ing. Dr. Reinhard Perfler

University of Natural Resources and Applied Life Sciences, Vienna, Austria

Execution

Mr. Emmanuel Kwandahor

Physical Plant Department, Valley View University, Accra

Ecological Engineering Society, IÖV

Valley View University, VVU

Financed by

Federal Ministry for the Environment, Nature Conservation and Nuclear Safety, Germany

General Background

Water is a precious resource. It develops from surface water (e.g. rain, river, lake) which has infiltrated into the ground and moved very slowly throughout different permeable ground layers. Its quality is usually higher than surface water as the soil layers provide better protection against pollution. Groundwater contains minerals resulting from the dissolution of sediments and rocks. Some of these minerals (e.g. calcium or magnesium) are supporting healthy nutrition.

Water at VVU

At VVU, the groundwater table is about 40 m below surface. The water is very likely to be originating from the Akwapim-Togo range and has remained in the ground for a time longer than 20 years. VVU is located at the border of two main geological formations, the Togo series and the Acidic Dahomeyan. In this area, the borehole yields are generally ranging from 1 - 3 m^3/h.

At VVU there are 3 boreholes located near the Baobab Centre. Borehole n°3 is rarely used because it delivers turbid water. The groundwater abstraction amounts to an average of 25 m^3 per day, mostly out of borehole n°1. In comparison, the provision from Oyibi Water Supply Scheme was 40 m^3 per day during the period 2006-2009.

Analyses of the groundwater quality have indicated iron concentration occasionally exceeding Ghana drinking water requirements. High iron concentration imparts colour and can lead to the formation of iron deposit in the supply system. The water can also be very turbid and is also slightly acid with a pH ranging from 5,4 to 5,9.

Filter characteristics for Evers Easy Filtration® 400

Treatment volume capacity	0,3-1,2 m^3/h
Filtration velocity for treatment	3-10 m/h
Diameter of filter column	0,4 m
Height of filter material layer	1,2 m
Total height of filter column	2,6 m
Volume of container for backwashing	0,6 m^3
Minimum pressure head for backwashing	3,0 m
Backwash time	8 min
Filtration velocity for backwashing	36 m/h
Filter material grain size	0,8-1,6 mm
Filter material density	1,40-1,45 g/cm^3

Borehole drilling at VVU (January 2008)

Well head at borehole n°1

Turbid water, borehole n°3

Water treatment at VVU

A groundwater treatment has been installed near the main reservoir inside a roofed container. The treatment goals are to reduce the turbidity of groundwater and the concentration of iron to meet the drinking water requirements for these two parameters.

Evers Easy Filtration® 400

Filtration material

Besides treating groundwater from abstraction wells, the treatment units can also be used to filter rainwater stored in the underground storage tank of the Baobab Centre.

Water monitoring at VVU

Groundwater is a resource, which should be used wisely and carefully to avoid over exploitation and pollution. Measurement devices were installed in the boreholes to monitor the evolution of the groundwater level and the water quality.

Location of boreholes

Poster: Sachet Water Production at Valley View University

CLIMATE CHANGE
INITIATIVE PROJECT
2009/2010

Water treatment

Filtering, monitoring water quality and production of sachet water

Design and Supervision

Dipl.-Ing. Laurent Richard M.Sc.
Dipl.-Ing.Dr. Reinhard Perfler
University of Natural Resources and Applied Life Sciences, Vienna, Austria

Execution

Mr. Emmanuel Kwandahor
Physical Plant, Valley View University, Accra

Ecological Engineering Society, IÖV
Valley View University, VVU

Financed by
Federal Ministry for the Environment, Nature Conservation and Nuclear Safety, Germany

General background of the use of packed drinking water

At Valley View University, as in most areas of Ghana, drinking water is not taken from the tap but from bottles or sachets. Sachets are small plastic containments made of high density polyethylene, which are filled with drinking water.

Within the actual project it was decided to implement a sachet water production at VVU. The revenues shall be used for ecological development at VVU.

Water treatment for sachet water production

Water from VVU supply system, a mixture of VVU groundwater and Oyibi water, is used for the sachet water production. In the future, the production should be based solely on groundwater abstraction at VVU campus. The aim is to produce drinking water for the whole campus. The production target is 4,5 m³/day.

Actual measurements show that the physicochemical quality of the water is complying with the drinking water requirements. However, the water must be disinfected.

The sachet water factory at VVU uses a similar production process as other sachet water factories. It consists of a storage tank, a distribution system (pipes), a decentralized water treatment system (filters and UV-C disinfection unit), and a packaging system. The main feature of the treatment at VVU is the use of a low-pressure membrane filter system.

In addition, the production will follow a strict quality management procedure to ensure safe and best water quality. The plastic waste originating from sachet water production and use at VVU will be collected for recycling.

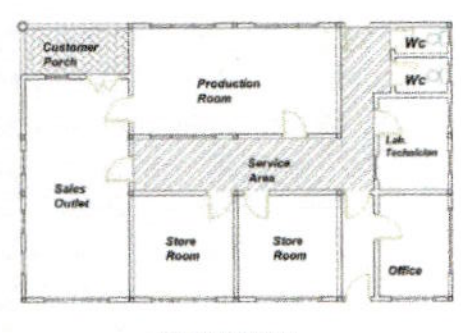

Production factory, Floor Plan

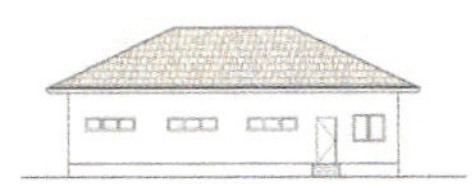

Production factory, Elevation

The SkyHydrant™ unit only requires a low amount of chemicals and energy for water processing. It does not remove salt or dissolved chemicals and minerals from the water, which are important to support healthy nutrition. However, it removes turbidity and particles larger than 0,1 µm. It also removes significantly most of the biological contaminants and pathogens including bacteria, viruses, protozoa, cysts and other parasites. The remaining biological contaminants are inactivated during the subsequent UV-C disinfection.

The SkyHydrant™ Unit comprises a single MEMCOR® membrane sub-module in a low-pressure housing. The unit is suitable for operation under low positive or negative head pressure. Fresh water flows along the length of the hollow PVDF fibres before being forced through the walls of the fibre to produce a filtrate virtually free of suspended solids .

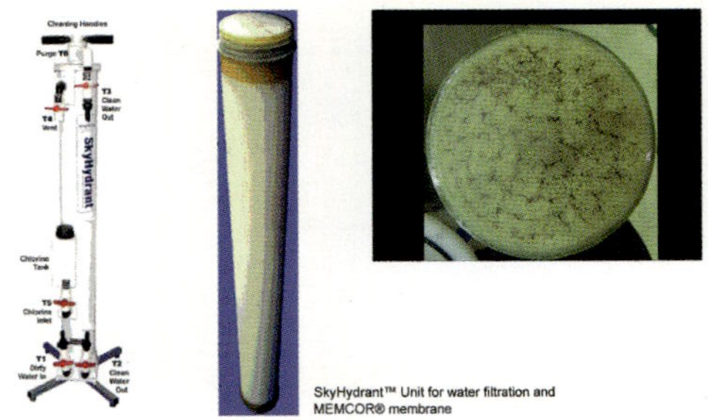

SkyHydrant™ Unit for water filtration and MEMCOR® membrane

The unit is designed for long-term use. It can have a life span of 5 to 10 years when properly maintained. The operation and maintenance (manual and chemical cleaning) is simple and requires no specific tools or skilled specialists.

Compared to the commonly used polypropylene cartridge filters (filtration down to 0,5 µm), the SkyHydrant™ unit offers better filtration performance (filtration down to 0,1 µm). It also produces less waste as the membrane can be cleaned and used several years, whereas the polypropylene cartridge filters have to be replaced frequently.

Poster: Rainwater Storage Underground Roads at Valley View University

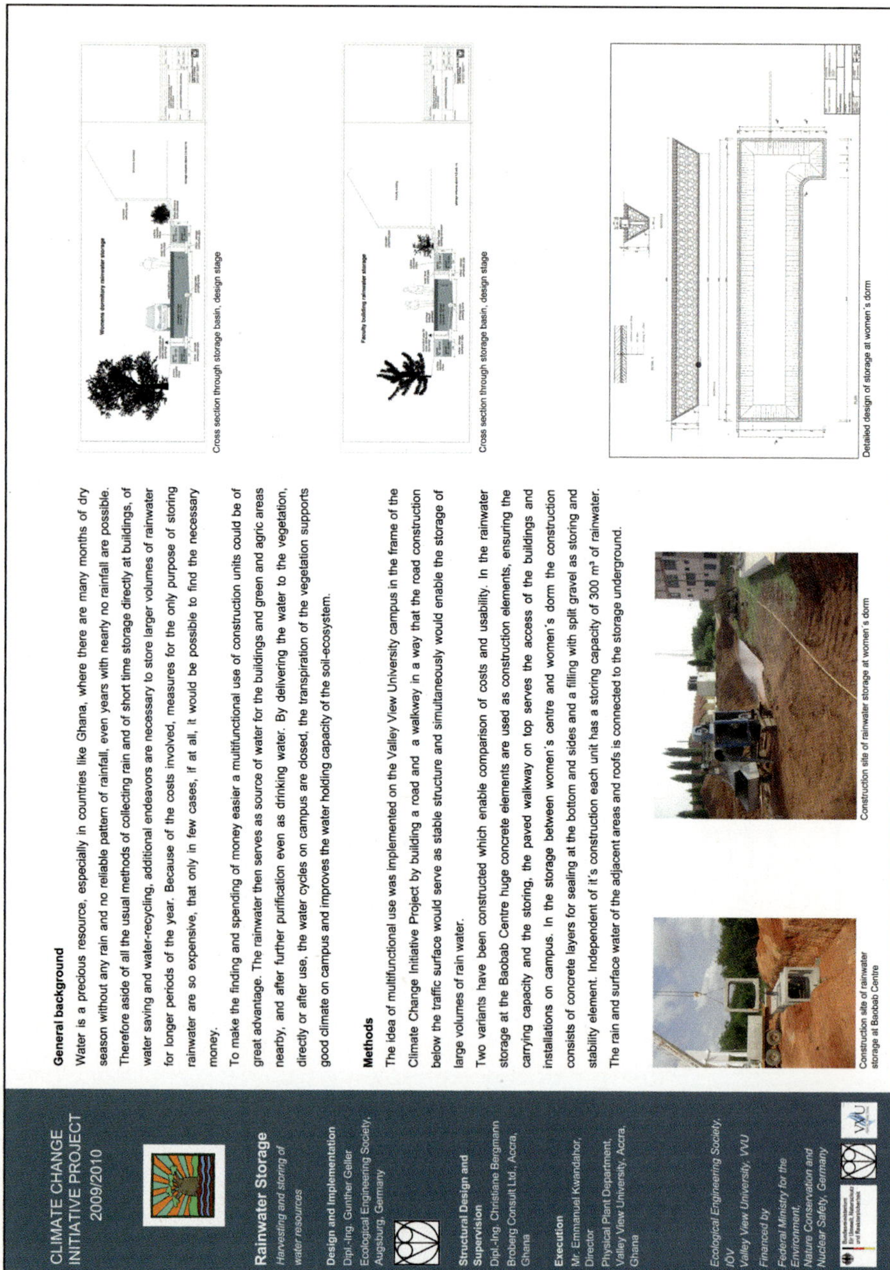

EU-ECOCITY-Project: Materials

	CRITERIA	INDICATORS
CON-TEXT	location of settlement	basic urban supply infrastructure — *potential, attractiveness and accessibility of means for satisfying basic needs* coping with land demand — *ratio of brownfield, inner city and greenfield development areas per total planning area*
URBAN STRUCTURE	building density	area density - floor area per land area
	mix of uses	share of mixed-use areas – *share of gross floor area with a mix of residential and non-residential use* accessibility of basic facilities – *proximity to groceries; kindergarten; primary school; pubs*
	public spaces *size and quality*	conviviality index - *quantity of public spaces with potential for conviviality* public space quality - *liveability (active frontages + diversity of uses), accessibility, legibility, safety, connectivity, urban comfort*
	landscaping area *accessibility and surface quality*	accessibility of green areas – *share of inhabitants living near public green areas* eco-quality of outdoor areas – *e.g. artificial cover, cut / uncut grass, trees, permanent / temporary water bodies, roof and facade planting*
TRANSPORT	infrastructure for private travel	application of transport concepts for the *reduction of car traffic* length of road network *per workday population* length of cycle network *per workday population*
	accessibility of public transport	public transport coverage - *percentage of floor area within a radius of 300 m or 150 m of public transport stops*
	quiet *noise from routes of transport*	daytime traffic noise exposure night time traffic noise exposure *share of inhabitants exposed to noise above limits*
	parking provision	accessibility of public transport vs. private car
ENERGY FLOWS	energy demand	annual energy demand – *for heating, cooling and other purposes* *peak power demand per floor area*
	energy efficiency	compactness of the structure ratio of solar oriented buildings thermal insulation level
	greenhouse gas emissions	share of renewable energy sources global warming potential per MWh - *CO_2-equivalent of non-renewable energy production*
MATERIAL FLOWS	building materials	reduction of the demand for building materials use of renewable, recycled and/or locally-sourced building materials
	soil movement	share of soil re-used on site
	water management	concepts for water management – *measures to minimise primary water consumption*
SOCIO-ECONOMIC ISSUES	social infrastructure and mix	index of social infrastructure – *availability/existence of social institutions; measures for social diversity and integration*
	economic infrastructure	index of economic infrastructure – *quality of local economic development plan*
	labour related issues	index of labour related issues – *availability and diversity of jobs, (correspondence to social mix and the economic context)*
	profitability	index of profitability – *cost/return ratio*
PRO-CESSES	integrated planning	multidisciplinary planning team - *disciplines and institutions integrated* iterative process – *number of optimisation loops* planning with scenarios – *number and content of scenarios*
	community involvement	index of community involvement – *quality of the participation process*

Fig. A.1 ECOCITY evaluation scheme: criteria and indicators (Gaffron et al. 2008, 71)

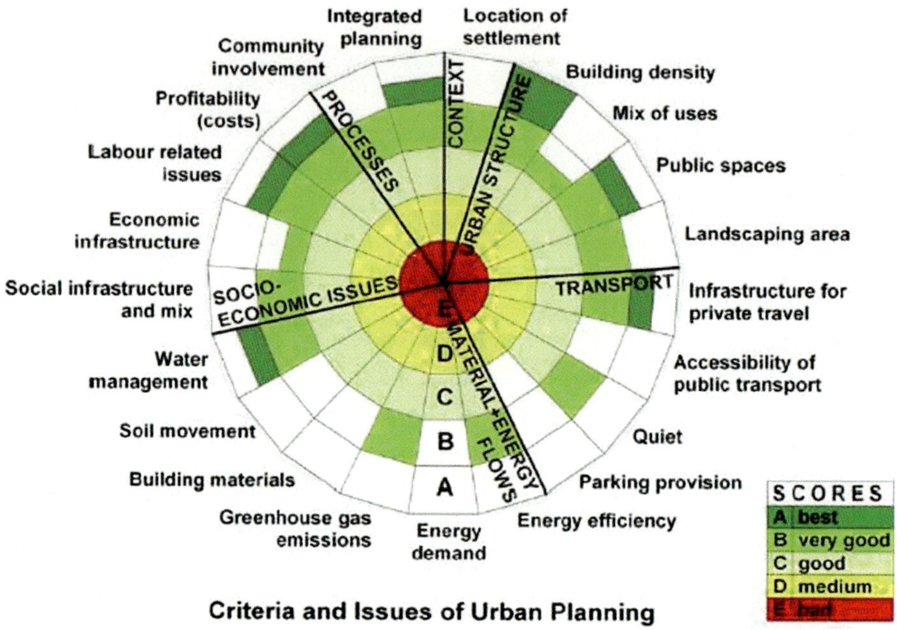

Fig. A.2 Evaluation compass: criteria and issues of ECOCITY planning (Gaffron et al. 2008, 73)

Bibliography

Abbiw, D. K. 1990. The useful plants of Ghana. - Royal Botanic Gardens, Kew, Richmond, Surrey.

Abbiw, D. K. no year. Endangered medicinal plants in the middle and southern sector of Ghana. – Traditional and Alternative Medicine Directorate, Ministry of Health, P.O. Box MB-44, Accra.

African Development Bank/Organisation for Economic Co-operation and Development (AfDB/OECD), 2007. African Economic Outlook 2007, http://www.oecd.org/dataoecd/26/51/38562673.pdf

Awadzi, T., Bredwa-Mensah, Y., Breuning-Madsen, H., Boateng, E., A. 2001. Scientific evaluation of the agricultural experiments at Frederiksgave, the Royal Danish Plantation on the Gold Coast, Ghana. Danish Journal of Geography 2001, 101: 33–40.

Baldwin, Chr. 1998. Calling the Circle. The First and Future Culture. Bantam Books, New York: 245.

Baker, H. G. 1962. The ecological study of vegetation in Ghana. In Agriculture and Land Use in Ghana, Wills, J. B. P., Ed. Oxford University Press: London, Accra, New York: 151–159.

Bianchi, F. J. J. A. Booij, C. J. H., Tscharntke, T. 2006. Sustainable pest regulation in agricultural landscapes: a review on landscape composition, biodiversity and natural pest control. Proceedings of the Royal Society B: Biological Sciences: 273, (1595): 1715–1727.

Brammer, H., 1962. Soils: In Agriculture and Land Use in Ghana, Wills, J. B. P., Ed. Oxford University Press: London, Accra, New York; 160–169.

Braungart, M., McDonough, W. 2003. Cradle to cradle. Remaking the way we make things. North Point Press, New York: 193.

Braungart, M., McDonough, W. 2003. Einfach intelligent produzieren. Berliner Taschenbuch Verlag, Berlin: 236.

Brown, J., Isaacs, D. 2005. The World Café. Shaping Our Futures Through Conversations That Matter. Berret-Koehler, San Francisco: 242.

Cockerill, K. & Carp, J. 2009. Leveraging opportunities for campus sustainability: a case study of water resources. Sustainability: Science, Practice & Policy. 5(2): 28–37.

Engel, A., Korf, B. 2005. Negotiation and mediation techniques for natural resource management. – FAO. http://www.fao.org/docrep/008/a0032e/a0032e00.htm

Fuchs, O., Schleifnecker, Th. 2001. Handbuch ökologische Siedlungsentwicklung. Konzepte zur Realisierung zukunftsfähiger Bauweisen. Erich Schmidt, Berlin: 300.

Gaffron, P., Huismans, G., Skala, F. 2005. Ecocity Book I. A better place to live. Facultas Verlags- und Buchhandels AG, Vienna: 108.

Gaffron, P., Huismans, G., Skala, F. 2008. Ecocity Book II. How to make it Happen. Facultas Verlags- und Buchhandels AG, Vienna: 84.

Gagel, D. 1995. Action Research and Small Enterprise Promotion. Methods of participatory project planning and implementation in development cooperation. GTZ, Eschborn: 237.

G. Geller and D. Glücklich (eds.), *Sustainable Rural and Urban Ecosystems: Design, Implementation and Operation*, DOI 10.1007/978-3-642-28261-4,

Gairola, S., Noresah, M. S. 2010. Emerging trend of urban green space research and the implications for safeguarding biodiversity: a viewpoint. Nature and Science 2010, 8, (7): 43–49.

Geller, G., Berger, W., Fries, N., Germer, J., Glücklich, D., Höner, G., Laryea, S., Sauerborn, J., 2006. Ecological Development of Settlements – the case study Valley View University, Accra, Ghana. Research & Development in Sub-Saran Africa, 12: 2–5.

Geller, G., 2006. Ökologische Kreislaufwirtschaft beim Ausbau des Campus der Valley View University in Accra, Ghana. 1. Aachener Kongress Dezentrale Infrastruktur am 17. und 18. Oktober 2006. Gewässerschutz, Wasser, Abwasser 204: 23/1-23/10.

Girardet, H. 2008. Creating Sustainable Cities. Green Books, Foxhole, Dartington for The Schumacher Society Foxhole, Dartington: 77.

Glücklich, D. (Hrsg.) 2004. Ökologisches Bauen. Von Grundlagen zu Gesamtkonzepten. DVA, München: 246.

Glücklich, D. 2002 (Hrsg.). New Old Ecology? Ökologisches Bauen – Bilanz und Perspektiven. Universitätsverlag Weimar, Weimar: 190.

Glücklich, D. (Hrsg.) 2010. Werkstattkoffer Ökologisches Bauen.

Glücklich, D. (Hrsg.) 2010. Master-Fernstudiengang Architektur und Umwelt. Hochschule Wismar, WINGS. Studienbegleitende Unterlagen.

Grebe (Hrsg.) 1977. Vorentwurf zum Landschaftsplan Köln als Beitrag zum Flächennutzungsplan. Planungsbüro Grebe, Nürnberg: 142.

GreenKeys Project Team 2008: Greenkeys@Your City. A Guide for Urban Green Quality. IOER Leibniz Institute of Ecological and Regional Development. Dresden: 105, CD included.

Grenier, Louise 1998. Working with Indigenous Knowledge. A Guide for Researchers. International Development Research Centre, Ottawa: 115.

Gyekye, K. 1996. African Cultural Values. An Introduction. Sankofa Publishing Company. Accra: 194.

Hall, J. B., Swaine M. D. 1976. Classification and ecology of closed-canopy forest in Ghana. – Journal of Ecology 64 (3): 913–951.

Hall, J. B., Swaine, M. D. 1981. Distribution and ecology of vascular plants in a tropical rain forest. - Forest vegetation in Ghana. Dr. W. Junk Publishers: The Hague, 1981; Vol. 1: 383.

Harum, F., Moestrup, S. 2010. Manual – Tree Nursery Establishment for Small Holder Farmers. University of Copenhagen, Burung Indonesia, Dansk Ornitologisk Forening. Forest & Landscape Denmark: 16. http://en.sl.life.ku.dk/Forskning/DevelopmentEnvironment/Publications.aspx

House, S. J., Reed, R. A. 1997. Emergency Water Sources: Guidelines for selection and treatment. Water Engineering and Development Centre (WEDC). Loughborough. http://www.who.or.id/eng/contents/aceh/wsh/books/ews/ews.htm

Ingenieurökologische Vereinigung IÖV (Ed) 2010. Ecological Cycle Management at Valley View University in Accra, Ghana. Final Report of the BMBF-Research Project 02WD 0472–76. Augsburg: 110+Appendix

IAIA. 2005. Biodiversity in Impact Assessment. IAIA Special Publications Series 3. – IAIA: 1–4. http://www.iaia.org/ publicdocuments/special-publications/SP3.pdf

IAIA, B. S. 2010. Biodiversity - Key Citations Series. - International Association of Impact Assessment. http://www.iaia.org/publicdocuments/Activity_Resources/Key_Citations/Key%20Citations_Biodiversity%2010%20Feb.pdf

Jeník, J., Hall, J. B. 1976. Plant communities of the Accra Plains, Ghana. – Folia Geobotanica & Phytotaxonomica 11 (2): 163–212.

Kellogg, s., Pettigrew, S. 2008. Toolbox for Sustainable City Living. A Do-It-Ourselves Guide. South End Press. Cambridge, MA:242.

Krusche P. und M., Althaus, D., Gabriel, I. 1982. Ökologisches Bauen. Bauverlag Wiesbaden, Berlin: 360.

Lane, D. A. 1962. The forest vegetation. In Agriculture and Land Use in Ghana, Wills, J. B. P., Ed. Oxford University Press: London, Accra, New York: 160–169.

Longman, KA. 1993. Tropical trees: propagation and planting manuals. Volume 1: Rooting cuttings of tropical trees. Commonwealth Science Council, London, UK. http://www.fao.org/DOCREP/006/AD231E/AD231E00.HTM

Longman, KA. 1995. Tropical trees: propagation and planting manuals. Volume 4: Preparing to plant tropical trees. Commonwealth Science Council, London, UK. http://www.fao.org/DOCREP/006/AD229E/AD229E00.HTM

Longman, KA. 1998. Tropical trees: propagation and planting manuals. Volume 3: Growing good tropical trees for planting. Commonwealth Science Council, London, UK. http://www.fao.org/DOCREP/006/AD228E/AD228E00.HTM

Longman, KA. 2003. Tropical trees: propagation and planting manuals. Volume 2: Raising seedlings of tropical trees. Commonwealth Science Council, London, UK. http://www.fao.org/DOCREP/006/AD230E/AD230E00.HTM

Lovell, S. T.; Johnston, D. M. 2009. Creating multifunctional landscapes: how can the field of ecology inform the design of the landscape? Frontiers in Ecology and the Environment 2009, 7, (4): 212–220.

Ludema, J., Whitney, D., Mohr, B., Griffin, Th. 2003. The Appreciative Inquiry Summit. A Practioner´s Guide for Leading Large Group Change. Berret-Koehler, San Francisco: 310.

Lyle, J. 1999. Design for Human Ecosystems: Landscape, Land Use, and Natural Resources. Island Press: 288.

Kramer, R. 2007. Leading Change Through Action Learning. The Public Manager, 36 (3): 38–44.

Krämer, 2007. Ecological development of the Valley View University, Accra, Ghana. Diploma thesis, Rheinmain University of Applied Sciences

Macy, J. 1991. Mutual Causality in Buddhism and General Systems Theory. The Dharma of Natural Systems. State University of New York Press, New York: 236.

Macy, J., Brown, M.Y. 2003. Coming Back to Life: Practices to Reconnect Our Lives, Our World. Jungfermann, Paderborn Island: 220

Macy, J., Brown, M.Y. 1999. Die Reise ins lebendige Leben. Strategien zum Aufbau einer zukunftsfähigen Welt. Ein Handbuch. New Society Publishers, Gabriola Island: 259.

Marzluff, J., Shulenberger, E., Endlicher, W., Alberti, M., Bradley, G., Ryan, C., ZumBrunnen, C., Simon, U. 2008. Urban Ecology – An International Perspective on the Interaction Between Humans and Nature. Springer: New York: 808.

Mitsch, W. J. 1996. Ecological Engineering: A new paradigm for engineers and ecologists. In: Schulze, P.C. (Eds) Engineering within Ecological Constrraints. National Academy Press. Washington, D.C.: 114–132.

Münch E. v.; Schöpe A.; Rüd S., 2009. Ecosan – Recycling oriented wastewater management and sanitation systems. GTZ. http://www.gtz.de/de/dokumente/en-ecosan-topicsheet-2009.pdf

Odum, H. T. 1962. Man in the Ecosystem. In: Proceedings Lockwood Conference on the Suburban Forest and Ecology. Bull. Conn. Agr. Station 652. Storrs. CT:57–75.

Odum, E. P. 1971. Fundamentals of Ecology. Saunders College Publishing. Philadelphia: 574

Owusua, G. 2010. Social effects of poor sanitation and waste management on poor urban communities: a neighborhood-specific study of Sabon Zongo, Accra – Journal of Urbanism: International Research on Placemaking and Urban Sustainability 3: 145–160.

Owusu-Mensa, K. 2009. Three decades of Ghana's Valley View University. – Manuscript, Valley View University

Palutec, 2008. Ökologische Kreislaufwirtschaft an der Valley View University Accra (Ghana) – Teilprojekt 4 – Abschlussbericht der Palutec GmbH (BMBF Verbundvorhaben, Förderkennzeichen 02WD0475)

Perfecto, I., Vandermeer, J., Wright, A. 2009. Nature's Matrix – Linking Agriculture, Conservation and Food Sovereignty. Earthscan: London: 242.

Reason, P., Bradbury, H. 2005. Handbook of Action Research, 2nd Edition. Sage Pubn Inc, London: 362.

Register, R. 2006. Ecocities. Rebuilding Cities in Balance with Nature. New Society Publishers, Gabriola Island, BC, Canada: 375.

Ripl, W. 2003. Water – the bloodstream of the biosphere. In: Falkenmark, M. & Folke, C.: Freshwater and welfare fragility – syndromes, vulnerability and challenges. Freshwater Special Issue. Phil. Trans. R. Soc. London. B 358: 1921–1934.

Roseland, M. 2006. Towards Sustainable Communities. Resources for citizens and their governments. New Society Publishers, Gabriola Island BC, Canada and Stony Creek CT, USA:

Scharmer, C.O. 2007. U-Theory. Leading from the Future as it Emerges. The Social Technology of Presencing. SoL, Cambridge MA: 533.

Sharp, L. 2009. Higher Education: the quest for the sustainable campus. Sustainability: Science, Practice & Policy. 5(1): 1–8.

Smet J., van Wijk C. 2002. Small Community Water Supplies: Technology, people and partnership. Delft, The Netherlands, IRC International Water and Sanitation Centre. (Technical Paper Series; no. 40). 585p. http://www.irc.nl/page/1917

Speer, A. und Partner 2009. A Manifesto for Sustainable Cities. Prestel, Munich:219.

Stepp, J.R., Jones, E.C., Pavao-Zuckerman, M., Casagrande, D., Zarger, R. K. 2003. Remarkable Properties of Human Ecosystems. Conservation Ecology 7(3):11. (online)

Thomas T. H., Martinson D.B., 2007. Roofwater Harvesting: A Handbook for Practitioners. Delft, The Netherlands, IRC International Water and Sanitation Centre. (Technical Paper Series; no. 49). 160p. http://www.irc.nl/page/37471

Tilley E. et al, 2008. Compendium of Sanitation Systems and Technologies. Swiss Federal Institute of Aquatic Science and Technology (Eawag). Dübendorf, Switzerland. http://www.eawag.ch/forschung/sandec/publikationen/compendium_e/index_EN

Tomåsek, W. 1979. Die Stadt als Ökosystem. Überlegungen zum Vorentwurf Landschaftsplan Köln–Landschaft+Stadt. 11: 51–60.

UNEP. 2002. Rainwater Harvesting And Utilisation: An Environmentally Sound Approach for Sustainable Urban Water Management: An Introductory Guide for Decision-Makers, http://www.unep.or.jp/Ietc/Publications/Urban/ UrbanEnv-2/index.asp

UNICEF. 2008. UNICEF Handbook on Water Quality. http://www.unicef.org/wash/files/WQ_Handbook_final_signed_16_April_2008.pdf

United Nations, Department of Economic and Social Affairs, Population Division 2008. World Urbanization Prospects: The 2007 Revision.

Upton, D., De Groot, P. 2008. Tropical trees: propagation and planting manuals. Volume 5: Planting and Establishment of Tropical Trees. Commonwealth Science Council, London, UK: 142.

Werner C.; Otterpohl R.; Jönsson H. 2004. 10 Recommendations for action from the Lübeck symposium on ecological sanitation, "Ecosan—closing the loop"—Proceedings of the 2nd International Symposium on ecological sanitation, 07–11 April 2003, Lübeck, Germany. http://www2.gtz.de/dokumente/bib/04-5004a.pdf

Whitney, D., Trosten-Bloom, A. 2003. The Power of Appreciative Inquiry: A Practice Guide to Positive Change. Berret-Koehler Publishers, San Francisco: 266.

WHO, 2006. Guidelines for drinking water quality, 3rd edition, Geneva, Switzerland, http://www.who.int/water_sanitation_health/dwq/gdwq3rev/en/

WHO/UNICEF Joint Monitoring Program on Water Supply and Sanitation, 2010. A snapshop of Drinking-water and Sanitation in the MDG region sub-Saharan Africa – 2010 Update. http://www.wssinfo.org/fileadmin/user_upload/resources/1284625893-Africa_snapshot_2010.pdf

Yankson, P. W. K. & Gough, K. V. 1999. The environmental impact of rapid urbanization in the peri-urban area of Accra, Ghana. - Danish Journal of Geography 99: 89–100.

Yeboah, I. E. A. 2000. Structural Adjustment and Emerging Urban Form in Accra, Ghana. - Africa Today 47: 61–89.

Index

G. Geller and D. Glücklich (eds.), *Sustainable Rural and Urban Ecosystems: Design, Implementation and Operation*, DOI 10.1007/978-3-642-28261-4,

Printed by Printforce, the Netherlands